Tourism, Cosmopolitanism and Global Citizenship

Certain types of tourism, such as volunteer tourism and student travel, have long been associated with global citizenship. To travel and to experience other societies and other cultures is linked with a cosmopolitan outlook, and also with the capacity to empathise and act ethically in relation to people in distant countries. In turn global citizenship – being a 'citizen of the world' – has become increasingly important both as a moral and political identity. Encouraged by employers, validated by universities, travel has become a marker of moral and intent for altruistic and ambitious youth with a mind to travel and the bank balance to facilitate it. The chapters in this volume explore the relationship between tourism, global citizenship and cosmopolitanism.

The chapters in this book were originally published as a special issue of *Tourism Recreation Research*.

Jim Butcher is a Reader in the School of Human and Life Sciences at Canterbury Christ Church University in the UK. He has written three books on the politics of tourism, and blogs at http://politicsoftourism.blogspot.co.uk/

Tourism, Cosmopolitanism and Global Citizenship

Edited by
Jim Butcher

LONDON AND NEW YORK

First published 2018
by Routledge
2 Park Square, Milton Park, Abingdon, Oxon, OX14 4RN, UK

and by Routledge
711 Third Avenue, New York, NY 10017, USA

Routledge is an imprint of the Taylor & Francis Group, an informa business

Introduction, Chapters 1–3 & 5–12 © 2018 Taylor & Francis.
Chapter 4 © 2017 Ruth Cheung Judge. Originally published as Open Access.

With the exception of Chapter 4, no part of this book may be reprinted or reproduced or utilised in any form or by any electronic, mechanical, or other means, now known or hereafter invented, including photocopying and recording, or in any information storage or retrieval system, without permission in writing from the publishers. For details on the rights for Chapter 4, please see the chapter's Open Access footnote.

Trademark notice: Product or corporate names may be trademarks or registered trademarks, and are used only for identification and explanation without intent to infringe.

British Library Cataloguing in Publication Data
A catalogue record for this book is available from the British Library

ISBN 13: 978-1-138-48308-8

Typeset in Myriad Pro
by RefineCatch Limited, Bungay, Suffolk

Publisher's Note
The publisher accepts responsibility for any inconsistencies that may have arisen during the conversion of this book from journal articles to book chapters, namely the possible inclusion of journal terminology.

Disclaimer
Every effort has been made to contact copyright holders for their permission to reprint material in this book. The publishers would be grateful to hear from any copyright holder who is not here acknowledged and will undertake to rectify any errors or omissions in future editions of this book.

Contents

Citation Information   vii
Notes on Contributors   ix

Introduction: Tourism, cosmopolitanism and global citizenship   1
Jim Butcher

1. Citizenship, global citizenship and volunteer tourism: a critical analysis   3
Jim Butcher

2. A rite of passage? Exploring youth transformation and global citizenry in the study abroad experience   13
Simone Grabowski, Stephen Wearing, Kevin Lyons, Michael Tarrant and Adam Landon

3. Cosmopolitan empathy in volunteer tourism: a psychosocial perspective   24
Émilie Crossley

4. Class and global citizenship: perspectives from non-elite young people's participation in volunteer tourism   38
Ruth Cheung Judge

5. 'FEEL IT': moral cosmopolitans and the politics of the sensed in tourism   50
João Afonso Baptista

6. Cosmopolitan education, travel and mobilities to Washington, DC   62
Felix Schubert and Kevin Hannam

7. Producing science and global citizenship? Volunteer tourism and conservation in Belize   73
Noella J. Gray, Alexandra Meeker, Sarah Ravensbergen, Amy Kipp and Jocelyn Faulkner

8. Volunteer tourism in Romania as/for global citizenship   86
Cori Jakubiak and Iulia Iordache-Bryant

9. Mediating global citizenry: a study of facilitator-educators at an Australian university   97
Tamara Young, Joanne Hanley and Kevin Daniel Lyons

10. Educating tourists for global citizenship: a microfinance tourism providers' perspective   109
Giang Thi Phi, Michelle Whitford, Dianne Dredge and Sacha Reid

11. The limits of cosmopolitanism: exchanges of knowledge in a Guatemalan volunteer programme   122
Rebecca L. Nelson

12. There's a troll on the information bridge! An exploratory study of deviant online behaviour impacts on tourism cosmopolitanism   132
Aaron Tham and Mingzhong Wang

Index   147

Citation Information

The chapters in this book were originally published in *Tourism Recreation Research*, volume 42, issue 2 (2017). When citing this material, please use the original page numbering for each article, as follows:

Editorial
Volunteer tourism, cosmopolitanism and global citizenship
Jim Butcher
Tourism Recreation Research, volume 42, issue 2 (2017), pp. 127–128

Chapter 1
Citizenship, global citizenship and volunteer tourism: a critical analysis
Jim Butcher
Tourism Recreation Research, volume 42, issue 2 (2017), pp. 129–138

Chapter 2
A rite of passage? Exploring youth transformation and global citizenry in the study abroad experience
Simone Grabowski, Stephen Wearing, Kevin Lyons, Michael Tarrant and Adam Landon
Tourism Recreation Research, volume 42, issue 2 (2017), pp. 139–149

Chapter 3
Cosmopolitan empathy in volunteer tourism: a psychosocial perspective
Émilie Crossley
Tourism Recreation Research, volume 42, issue 2 (2017), pp. 150–163

Chapter 4
Class and global citizenship: perspectives from non-elite young people's participation in volunteer tourism
Ruth Cheung Judge
Tourism Recreation Research, volume 42, issue 2 (2017), pp. 164–175

Chapter 5
'FEEL IT': moral cosmopolitans and the politics of the sensed in tourism
João Afonso Baptista
Tourism Recreation Research, volume 42, issue 2 (2017), pp. 176–187

Chapter 6
Cosmopolitan education, travel and mobilities to Washington, DC
Felix Schubert and Kevin Hannam
Tourism Recreation Research, volume 42, issue 2 (2017), pp. 188–198

Chapter 7
Producing science and global citizenship? Volunteer tourism and conservation in Belize
Noella J. Gray, Alexandra Meeker, Sarah Ravensbergen, Amy Kipp and Jocelyn Faulkner
Tourism Recreation Research, volume 42, issue 2 (2017), pp. 199–211

Chapter 8
Volunteer tourism in Romania as/for global citizenship
Cori Jakubiak and Iulia Iordache-Bryant
Tourism Recreation Research, volume 42, issue 2 (2017), pp. 212–222

Chapter 9
Mediating global citizenry: a study of facilitator-educators at an Australian university
Tamara Young, Joanne Hanley and Kevin Daniel Lyons
Tourism Recreation Research, volume 42, issue 2 (2017), pp. 223–234

Chapter 10
Educating tourists for global citizenship: a microfinance tourism providers' perspective
Giang Thi Phi, Michelle Whitford, Dianne Dredge and Sacha Reid
Tourism Recreation Research, volume 42, issue 2 (2017), pp. 235–247

Chapter 11
The limits of cosmopolitanism: exchanges of knowledge in a Guatemalan volunteer programme
Rebecca L. Nelson
Tourism Recreation Research, volume 42, issue 2 (2017), pp. 248–257

Chapter 12
There's a troll on the information bridge! An exploratory study of deviant online behaviour impacts on tourism cosmopolitanism
Aaron Tham and Mingzhong Wang
Tourism Recreation Research, volume 42, issue 2 (2017), pp. 258–272

For any permission-related enquiries please visit:
http://www.tandfonline.com/page/help/permissions

Notes on Contributors

João Afonso Baptista is a Research Fellow at the Institute of Ethnology, the University of Hamburg, Germany, and at the Institute of Social Sciences, the University of Lisbon, Portugal.

Jim Butcher is a Reader in the School of Human and Life Sciences at Canterbury Christ Church University in the UK. He has written three books on the politics of tourism, and blogs at http://politicsoftourism.blogspot.co.uk/

Émilie Crossley is a recent graduate of Cardiff University, UK, whose work explores tourist subjectivity from the perspective of psychosocial studies and critical psychology. She is currently based at Otago Polytechnic in New Zealand.

Dianne Dredge is Professor at the Department of Culture and Global Studies, Aalborg University, Copenhagen, Denmark.

Jocelyn Faulkner is a masters student in the Department of Geography at the University of Guelph, Ontario, Canada. She is interested in alternative models of volunteer tourism as expressions of global citizenship and cosmopolitanism.

Simone Grabowski is a Research Associate at the University of Technology Sydney, Australia.

Noella J. Gray is Assistant Professor in the Department of Geography at the University of Guelph, Ontario, Canada.

Joanne Hanley is a Research Manager in the Centre for English Language and Foundation Studies at the University of Newcastle, Australia. Her research expertise includes higher education, regional development, tourism studies and community sociology.

Kevin Hannam is Head of Tourism and Languages and Professor of Tourism Mobilities in the Business School at Edinburgh Napier University, Craiglockhart Campus, UK.

Iulia Iordache-Bryant worked as an English Language Teaching Fellow at Payap University in Chiang Mai, Thailand, after graduating from Grinnell College, USA, with a joint degree in Psychology and Russian. Her current area of research interest includes international education in Southeast Asia.

Cori Jakubiak is Assistant Professor of Education at Grinnell College, USA, where she teaches courses in educational foundations and applied linguistics.

Ruth Cheung Judge is a social and cultural geographer who completed her PhD in the UCL Geography Department, UK, where she now teaches.

Amy Kipp is a masters student in the Department of Geography at the University of Guelph, Ontario, Canada. Her thesis analyses volunteer tourism in relation to gender and geographies of care and global citizenship.

Adam Landon is a Postdoctoral Researcher in the Warnell School of Forestry and Natural Resources at the University of Georgia, USA.

Kevin Daniel Lyons is Professor of Tourism and Management in the Newcastle Business School, Australia. His research spans the fields of leisure, sports, recreation and tourism management, with a focus on volunteer tourism, youth travel and global citizenship.

Alexandra Meeker completed her MA in the Department of Geography at the University of Guelph, Ontario, Canada. Her thesis examined volunteer tourism as a form of neoliberal conservation in Belize.

Rebecca L. Nelson is the Executive Director of America Solidaria, USA.

NOTES ON CONTRIBUTORS

Giang Thi Phi is a PhD candidate at the Department of Tourism, Sport and Hotel Management, Griffith University, Australia.

Sarah Ravensbergen completed her MA in the Department of Geography at the University of Guelph, Ontario, Canada. Her thesis explored community perceptions of and responses to volunteer tourism in Belize.

Sacha Reid is a Senior Lecturer and Higher Degree Research Convenor at the Department of Tourism, Sport and Hotel Management, Griffith University, Australia.

Felix Schubert is currently a PhD candidate in the Business School at Edinburgh Napier University, UK.

Michael A. Tarrant is Josiah Meigs Distinguished Teaching Professor at the University of Georgia in Athens, GA, USA.

Aaron Tham is a Lecturer and Researcher in the School of Business, Faculty of Arts, Business and Law, University of the Sunshine Coast, Australia.

Mingzhong Wang joined USC in 2015 as a Lecturer in ICT. His research areas include large-scale distributed systems, data management, multiagent systems, as well as information security issues.

Stephen Wearing is a Conjoint Professor at the University of Newcastle (UoN), Australia.

Michelle Whitford is a Senior Lecturer and Program Director of the Bachelor of International Tourism and Hotel Management at the Department of Tourism, Sport and Hotel Management, Griffith University, Australia.

Tamara Young is a Senior Lecturer in Tourism Studies in the Newcastle Business School, Australia. Her core research interests are cultural tourism, tourist cultures, tourism education and student mobility.

INTRODUCTION

Tourism, cosmopolitanism and global citizenship

Up until the relatively recent growth and growing profile of ethical tourism niches, it was unusual to associate tourism with global citizenship or cosmopolitan ambition: leisure was leisure and citizenship was political. If you wanted to make a difference to the world, or to study it, holidays were not the time or place to do it.

Political and social identities have changed. Sociologists refer to the decline of 'grand narratives' of Left, Right and nation. Fukuyama famously referred to 'the end of history', an era in which the major contestation of the organisation of society has been substantially resolved (1992). Undoubtedly, the end of the Cold War brought into sharp relief the apparent exhaustion of erstwhile political identities through which individuals made sense of themselves in the wider world. The political realm increasingly became a question of technocratic management of *what is* rather than future-oriented ideologies relating to *what could be* (Furedi, 2013; Jacoby, 1999).

One result of this has been the rise of lifestyle-oriented attempts to act upon the world and its problems, outside of the political realm as previously understood. These attempts are often based around consumption and personal experience, and involve the elevation of pre-political virtues such as care and responsibility to others over distinctive political philosophies of market, state and democracy. The human and humane impulse to act in support of others, cut adrift from Politics (with a capital 'P'), becomes more a part of Giddens' life politics (or politics with a small 'p') (1994).

The rise of volunteer tourism is a good example of the trend Giddens noted. Its rise over the last 30 years has tended to associate tourism with wider moral, social and even political projects. Whilst others are the prospective benefactors, these are very much also projects of ethical selfhood: the forging of an ethical sense of self in a world in which the import of older political and moral parameters has diminished.

According to the influential website VolunTourism.org., the first ever use of the term 'voluntourism' (synonymous with volunteer tourism) was by the Nevada Board of Tourism as recently as 1988, who coined it to encourage volunteers to help in rural tourism projects (voluntourism.org: undated). David Clemmons, volunteer tourism entrepreneur and founder of VolunTourism.org, pointed out that the Google search engine had no hits for 'voluntourism' in 2000, but by 2010, the term yielded over 300,000 hits (cited in Vasquez, 2010). As I write, the figures are 556,000 hits for 'voluntourism', with over 2 million for the now more commonly used 'volunteer tourism'. Commercial volunteer tourism companies and ethical gap year organisers have boomed. Some non-governmental organisations have also adapted to the trend, offering visitor friendly trips to their development and conservation projects.

Volunteer tourism is closely associated with global citizenship in the literature and in commercial marketing (Butcher & Smith, 2015). This is also the case for ethical tourism more generally, although the link is more implicit. Whereas citizenship implies a relationship – legal and political – to the nation state, and via the state, to the world, global citizenship seeks moral solidarity beyond borders and operates through a global civil society of NGOs and ethical consumption. Hence, tourism can provide fertile ground for examining contemporary social and political identities, and raises novel questions for anyone interested in why and how people travel as tourists.

Global citizenship does not come with its own passport, or legal/political rights. It is, however, an important reference point for people's moral, social and political ambitions, connoting a cosmopolitan view of the world and a desire to act in support of others. Much debate has focused on the extent to which volunteer tourism does, or can, achieve its stated aims, or how it could be organised to succeed in this. Some have posed the question 'who benefits': the global citizen with an exciting and impressive portfolio of experiences, or the communities they seek to help. Others have sought to understand what the rise of volunteer tourism and other ethical niches indicates about contemporary culture.

It is in relation to volunteer tourism that tourism's association with cosmopolitanism and global citizenship is most developed. Stephen Wearing's *Volunteer tourism: Experiences that make a difference* (2001) paved the way for a wealth of writing from a variety of disciplines and perspectives. He and his co-writers have pushed the boundaries since. Mary Mostafanezhad, Jim Butcher and Peter Smith, Wanda Vrasti and others too have pursued a broadly social scientific approach, positioning volunteer tourism in the context of contemporary political, sociological and human geographical debates around neoliberalism, postcolonialism, the geographies of care and subjectivity. Others, such as Angela Benson, have focused on volunteer tourism's performance: what works, what is problematic and what is ethical (and on what basis).

Many others again have written on the topic with insight (see Wearing & McGehee, 2013 for a review). The ideas have been discussed regularly at many conferences and meetings over recent years, including those of the *American Association of Geographers* (AAG) and the *Association of*

Tourism and Leisure Studies (ATLAS). The mainstream press has also taken a strong interest, and articles about the efficacy of volunteer tourism, often based on tourists' own experience, are common.

Tourism as a spur to global citizenship is a theme that can be examined from many perspectives: sociological, psychological, political, anthropological and human geographical. It also has implications for business: its operation, marketing and also its ethical basis. The special issue reflects this diversity of perspectives, between and within papers. It includes papers that broach new conceptual and methodological ground. Others look at novel questions and cases. Taken together, they comprise a worthy addition, an important point of reference for scholars interested in ethical tourism and volunteer tourism as business or as a part of contemporary culture.

References

Butcher, J., & Smith, P. (2015). *Volunteer tourism: The lifestyle politics of international development*. London: Routledge.

Fukuyama, F. (1992). *The end of history and the last man*. London: Hamish Hamilton.

Furedi, F. (2013). *Authority: A sociological history*. Cambridge: Cambridge University Press.

Giddens, A. (1994). *Beyond left and right: The future of radical politics*. Oxford: Polity.

Jacoby, R. (1999). *The end of utopia: Politics and culture in an age of apathy*. New York, NY: Basic Books.

Vasquez, E. (2010, August 10). Do celebs like Jolie inspire voluntourism? *CNN*. Retrieved from http://edition.cnn.com/2010/TRAVEL/08/10/celebrity.humanitarian.travel/index.html

Wearing, S. (2001). *Volunteer tourism: Experiences that make a difference*. Wallingford: CABI.

Wearing, S., & McGehee, N. (2013). Volunteer tourism: A review. *Tourism Management, 38*, 120–130.

Jim Butcher

Citizenship, global citizenship and volunteer tourism: a critical analysis

Jim Butcher

ABSTRACT

This paper reflects on the association of volunteer tourism with global citizenship and argues that it involves outsourcing citizenship to 'the globe' in a manner unlikely to benefit global understanding or development politics. Volunteer tourism is strongly associated with global citizenship. Global citizenship, in turn, is associated with a better world. A key claim made about global citizenship is that it enables people to discharge their responsibilities to others in distant lands in an ethical way, less constrained by national interests. Yet global citizenship involves a reworking of the concept of citizenship not only spatially from nation to globe, but also politically from nation state and polity to non-governmental organisations and consumption (in this case, of tourism). The paper argues that in a number of ways the association of volunteer tourism with this geographically expanded but politically constricted notion of citizenship both reinforces a limited politics, and also limits the capacity of voluntourism to enlighten. By contrast, it is argued that a consideration of republican citizenship both clarifies these limits and suggests a more progressive rationale for volunteer travel.

Introduction

This paper develops a novel argument: a critique of the claims made for volunteer tourism as a promoter of a morally progressive global citizenship; and a restatement of the importance of republican citizenship as a framework through which to understand volunteer tourism's limits and potential as an enlightening human activity.

The context for this argument is the crisis in republican citizenship, and an attendant crisis of politics itself (Furedi, 2013; Jacoby, 1999), which have reinforced by default the moral and political claims associated with global citizenship. Put simply, in an age of post- and anti-politics (Swyngedouw, 2011), lifestyle can seem a more viable way to act upon the world (Giddens, 1994). It is only in this context that an activity considered until recently as politically facile as tourism can be part of a narrative of something as vital as citizenship (Butcher & Smith, 2015).

Yet there is a big, and neglected, contradiction. Global citizenship detaches citizenship from the polity and from the democratic or potentially democratic structures of the nation state (Parekh, 2003; Standish, 2012). Instead, this version of citizenship is enacted, or 'performed', through lifestyle and consumption (with tourism being a key example of this). The Arendtian (1958) view of an agonistic public sphere through which republican citizens can exercise their freedom collectively and in public through politics is replaced by essentially personal and private encounters and experiences (Butcher, 2015).

The argument developed here is that global citizenship is an inadequate moral and political framework for understanding the problems that volunteer tourists encounter, and one that may inhibit the potential in travel to prompt critical reflection and action on these problems. Given the commonplace explicit linking of volunteer tourism to global citizenship, the former provides much scope for looking critically at these issues.

The paper begins by clarifying citizenship and global citizenship as different and distinctive ways to consider the relationship of the individual to the social and political problems they encounter. It then considers how the literature (critically), and the providers (uncritically), consider volunteer tourism as a route to global citizenship. Drawing upon this framework, the paper then develops its critique, focusing on volunteer tourism's: capacity to address questions of power; cosmopolitan credentials and attraction as practical, 'doable' social action. Finally, it is argued that volunteer tourism 'outsources' the important political function of citizenship to 'the globe' in the name of global citizenship, but that this can be to the detriment of both the potential in volunteer travel and citizenship itself.

From the *polis* to global citizenship

The concept of citizenship originated with the *polis* in ancient Greece. Aristotle recognized man as a *zoon politicon* – a political animal. This feature of humanity was expressed through the *polis*, the ancient Greek city state. Citizenship progressed through its Roman conceptualisation, which involved a more developed legal relationship between citizen and state. The Italian City states of the Renaissance are also an important watershed in the development of citizenship, marking a shift away from subjects of a monarchy to citizens of a nation or city. In essence, citizenship involves the relationship between an individual and a political community, historically and culturally defined, within which social organisation is established and power legitimised and contested (Delanty, 2000; Faulks, 2000; Heater, 2004).

In modern society, citizenship developed in the context of the nation state. Citizens have rights within the state, sometimes inscribed in a constitution, as well as obligations under the law – the notion of a social contract is central. The civic republican conception of citizenship, championed by Hannah Arendt (2000, 1958), emphasises the individual operating in the public sphere, an active part of the political determinations that shape the society in which he or she lives.

Global citizenship is a very different model. Here identification with a 'global community' is emphasised above that as a citizen of a particular nation (Bianchi & Stephenson, 2014; Dower, 2003; Wilde, 2013). Global citizenship transcends geography or political borders and assumes that responsibilities and rights are or can be derived from being a 'citizen of the world'. This does not deny national citizenship, but the latter is often assumed to be more limited, morally as well as spatially (Dower, 2003).

The efficacy of global citizenship is premised upon the view that important political issues such as environmental damage, climate change and development are global in nature (Dower, 2003). That it may not be possible to address global-development-related issues from the perspective of nationally based politics is a common assumption (Dower, 2003). Issues are often presented as requiring private initiative (e.g. recycling, buying Fairtrade) linked to the globe (global poverty, globally unsustainable consumption), mediated through a global civil society of non-governmental organisations, globally oriented campaigns and also through ethical consumption (Delanty, 2000; Rhoads & Szelenyi, 2011; Standish, 2012).

Global citizenship as a concept has emerged principally through discussions about the role of education (Standish, 2012; see also various in Peters, Britton, & Blee, 2007). Advocates argue that children should learn about the world within a framework of global citizenship, and be encouraged to see themselves as having obligations towards environmental, human rights and development issues well beyond their own nation (Standish, 2012). This is especially the case in geography, but also true elsewhere in the curriculum (Standish, 2008). Today global citizenship has acquired a normative status in the education systems of Europe and North America and in much liberal thought. In the former case, US service learning aspires to promote global citizenship, and UK education from primary level through the Universities has adopted this outlook (Dill, 2013; Rhoads & Szelenyi, 2011).

According to one typical and influential definition, global citizenship means:

> enabling young people to develop the core competencies which allow them to actively engage with the world, and help to make it a more just and sustainable place. This is about a way of thinking and behaving. It is an outlook on life, a belief that we can make a difference. (Oxfam, in press)

School boards, educationalists and non-governmental organisations tend to follow a similar definition, and have become more involved in gap year projects and volunteer tourism type initiatives. Active engagement with 'the world' as opposed to national politics is emphasised. The references to 'a way of thinking and behaving' and 'an outlook on life' linked to global engagement suggest a broad moral orientation rather than a political or legal relationship. The 'belief that we can make a difference' in a new way is a typical, and very understandable, reaction to the stasis that seems to characterise the aforementioned post-political climate (Swyngedouw, 2011).

Global citizenship and volunteer tourism: the literature

Global citizenship, then, suggests a less-partial and less-bounded view of the world, and this corresponds to the lived travel experience of the mobile middle classes who comprise the bulk of the market for ethical tourism niches (Mowforth & Munt, 2015). Ethical tourism (in contrast to mass tourism) has long been linked with global citizenship *implicitly* (Krippendorf, 1987). More recently that link has become *explicit* and theorised, particular in relation to 'volunteer tourism', 'philanthropy tourism' and a variety of other niches associated with ethical travel (Lyons, Hanley, Wearing, & Neil, 2012; Novelli, 2004; Palacios, 2010; Phi, Dredge, & Whitford, 2013). Even the Gap Year – perhaps the closest thing western societies have to a rite of passage for middle class youth – is associated with, and occasionally even certificated for, promoting global citizenship (Heath, 2007; Simpson, 2005). The ethical traveller can, apparently, exercise their agency and morality in relation to the globe, directly and personally, through their travels.

Where this trend towards seeing leisure travel as linked to the moral and political project of global citizenship is at its most developed is in relation to volunteer tourism, or voluntourism. Here the link is explicit and strong (Lyons & Wearing, 2011; Lyons et al., 2012). There are variations, and varying emphases, in the associations made between volunteer tourism and global citizenship.

One view is that volunteer tourism can forge global citizenship by building long-term relationships and networks that promote activism in new social movements (McGehee, 2002; McGehee & Santos, 2005) particularly through promoting the understanding of other cultures (Crabtree, 2008; Devereux, 2008; Howes, 2008; McGehee, 2012). This point is made with emotional force in Generation NGO, a volume of highly personal accounts from young Canadian volunteers (Apale & Stam, 2011), some of whom pledge to act on the basis of lessons learned through their experience of other less economically developed societies. Notably, the activism in this case and others is not directed at transforming poverty into wealth, but at 'bringing home' the lessons learned abroad about how to live a more sustainable and co-operative life.

Palacios has argued strongly that volunteer tourism should drop any pretense to development and become more explicitly focused on promoting intercultural understanding and greater global awareness (2010). Effectively, volunteer tourism is held to have the potential to contribute to the forging of a global conscience and understanding key to the nurturing of ethical, global citizenship.

Global citizenship is also regarded as instrumental, as a credential achievable through travel that improves employment opportunities. Both Heath and Simpson have pointed out the importance of volunteer tourism in the building of a portfolio of experiences that can feature on a CV (Heath, 2007; Simpson, 2005).

There is also a substantial literature problematising volunteer tourism in terms of its neoliberal and/or neocolonial character (Mostafanezhad, 2013, 2014; Sin, 2009, 2010; Vrasti, 2012; Wearing, 2001, 2003; Wearing & Darcey, 2011; Wearing & Grabowski, 2011; Wearing & McGehee, 2013). The charge of neocolonialism in particular – that volunteer tourism can often draw upon and reinforce a narrative of northern benevolence meeting victimhood and gratitude in the global south, or the wealthy caring subject acting upon the impoverished object of their care – is relevant to this paper. Vrasti, for example, argues that the objectification of the Other in this way robs them of their humanity and their own agency, as well as ignoring the historical context of inequalities (2012). Neoliberalism – in this contact the marketisation of intercultural contact and culture itself – has also been argued to treat the hosts as *objects* of tourism, rather than as *subjects* in the context of an equal and authentic cultural exchange (Wearing, McDonald, & Ponting, 2005).

Although critical of much volunteer tourism, the literature also sees possibilities for challenging colonialism's legacy. Mostafanezhad argues that the objectification of the host in volunteer tourism can be challenged through images and accounts that disrupt the neoliberal narrative (Mostafanezhad, 2013). Wearing, McDonald and Ponting argue that decommodified forms of tourism (volunteer tourism having this potential) can challenge neoliberal assumptions and create spaces within which more authentic human interaction can develop (2005).

This paper will argue that the objectification of the host is better confronted by critiquing global citizenship claims and restating the value of republican citizenship. The former tends to abstract morality from the preeminent structures of political power and democracy, the nation, whereas the latter assumes individual and national sovereignty and agency.

Global citizenship and volunteer tourism: the providers

Volunteer tourism is a growing niche in many countries (ATLAS/TRAM, 2008; Smith & Holmes, 2009) and is organised by private companies, conservation and educational organisations, as well as non-governmental organisations (Broad, 2003; Raymond, 2008; Söderman & Snead, 2008). A survey of over 300 volunteer tourism organisations in 2008 concluded that the market caters to 1.6 million volunteer tourists per year, with a value of between £832 million and £1.3 billion ($1.7 billion–$2.6 billion) (ATLAS/TRAM, 2008). There is a focus on the 18–25 age range, those most likely to take a gap year (ATLAS/TRAM, 2008, p. 5; Jones, 2011, p. 535). The gap year itself is now a significant part of pre- and post-university life in the USA, Canada, Australia, New Zealand, UK and a number of other Europe countries (Lyons et al., 2012; Tomazos & Cooper, 2012).

The activities undertaken by volunteers are diverse. The range includes community work such as building a school or clinic, teaching English (Jakubiak, 2012) and conservation-based projects that involve scientific research or ecological restoration such as reforestation and habitat protection (Wearing, 2003). Typically volunteer projects involve linking community well-being and conservation in countries in the global South (Butcher & Smith, 2010).

As with the academic literature, in much of the promotion of volunteer tourism global citizenship themes are implicit. Many companies, non-governmental organisations and governmental initiatives have invoked global citizenship in their advertising explicitly too. For example, the non-profit Yanapuma Foundation offers a global awareness programme in Equador and the Galapagos Islands with the following statement on global citizenship:

> The concept of Global Citizenship encompasses sociocultural, political, economic and environmental factors as students experience at first hand the reality of being from the 'other side' of the development process. As such it implies critical and transformative elements as students develop their understanding on both social/political and personal experiential levels. The experience of immersion in a new context in combination with relevant academic support provides an intense learning environment that will transform both social/political awareness and personal awareness, informing future academic and professional development. (Yanapuma Foundation, in press)

A recent scheme promoted by a leading UK volunteer company asserts the link to employability from global citizenship, stating that volunteering overseas 'will help boost the employability skills and global citizenship of young adults'.

Politicians and commentators too have promoted the global citizenship benefits of a well-used Gap Year or a volunteer tourism project (Heath, 2007). For example, in 2009, the UK Department for Business, Innovation and Skills announced an initiative to assist students leaving university. In conjunction with expedition company Raleigh International (who arranged Prince William's volunteer placement in Chile in 2000), £500,000 of public funds was made available for 500 young people, under the age of 24 to travel to countries such as Borneo, Costa Rica, Nicaragua and India and participate in development and conservation projects.

Two years later the International Citizen Service was launched by the Prime Minister. The scheme operates through charities such as Raleigh International and Latitude Global Volunteering. International Citizen Service volunteers are expected to contribute to sustainable development abroad (including addressing the millennium development goals) and also to their own global citizenship via short unskilled volunteer placements.

Such initiatives are part of a wider orientation of politics towards volunteering, which has been promoted strongly by many governments (see various contributions to Paxton & Nash, 2002). In 2001, the UK Prime Minister announced that he intended 'to give more young people the chance of voluntary community service at home and abroad between school and university' (Chen, 2002). This contributed to a growth in political interest in the role of volunteering in citizenship. There have long been suggestions that volunteering may become a mandatory part of the school curriculum (Paxton & Nash, 2002), and it is commonplace for western Universities to integrate volunteering directly into the curriculum. In the UK, the notion of the 'Big Society', a centrepiece of the Conservative/Liberal Democrat coalition government of 2010–2015, was the latest in a line of high-profile government initiatives over the last 30 years designed to promote volunteering as a part of citizenship.

Schools, colleges and universities have also identified with the role of gap year projects and volunteer tourism in producing global citizens (Pearce & Coghlan, 2008). Educational institutions are where global citizenship is most in evidence. Careers advisors and geography admissions tutors often see it as a boon for employment opportunities in ethically attuned businesses (Standish, 2008). A growing number of universities even give formal academic accreditation to volunteer tourism trips and see it as an important part of creating global citizens (Jones, 2011). In North America there is a long-standing tradition of international service through all levels of education, and this tradition has also adapted to the more commercial and lifestyle orientation of volunteer tourism. Even the United States Peace Corps, as well as Voluntary Service Overseas (VSO) in the UK, whilst strongly resistant to the label volunteer tourism, have adapted to the trend, reducing both the need for qualifications and the length of service required on some programmes (Butcher & Smith, 2015).

Global citizenship: citizenship divorced from power and democracy

In contrast to the upbeat statements from volunteer tourism providers, and the normative value assigned to it in the academic literature on tourism, there are clear arguments against the moral efficacy of global citizenship. Sociologist Bikhu Parekh argues that: 'If global citizenship means being a citizen of the world, it is neither practicable nor desirable' (Parekh, 2003, p. 12). Such a citizenship is divorced from the actual institutions of politics that matter most – national governments. It is in the nation state that citizens can vote, or can strive for the vote, and through that alter the law, campaign for their rights and negotiate a social contract between state and individual (Parekh, 2003). The distant notion of a world state that would invalidate his opposition is also criticised as 'remote, bureaucratic, oppressive and culturally bland' (Parekh, 2003, p. 12). Global citizenship is citizenship divorced from power, and it is power that shapes the situation of the voluntoured.

Parekh's view is not to decry a knowledge of international issues, but to confront moral obligations towards others through a strengthened and agonistic relationship to national citizenship. He calls this being a 'globally oriented citizen' (Parekh, 2003), a national citizen who views their citizenship in the context of a political worldview.

Geographical education expert Alex Standish argues on a similar premise. He contends that global citizenship tends to bypass national politics in a world in which nations are the principal expressions of power and of democratic potential. Standish cites Heilman, who points out that 'cosmopolitan global citizenship [...] seeks to shift authority from the local to the national community to a world community that is a loose network of international organisations and subnational political actors not bound within a clear democratic constitutional framework' (cited in Standish, 2012, p. 66). Hence in bypassing national citizenship, global citizenship in a sense bypasses politics too.

The critiques of Standish and Parekh of the concept of global citizenship are apposite here. Global citizenship through volunteer tourism means citizenship carried out through private companies and non-governmental organisations. No one, bar shareholders, votes for the directors of companies. Non-governmental organisations and nonprofits are accountable to, at most, a self-selecting group of supporters. That is entirely appropriate for commercial companies and other organisations involved in the provision of volunteer tourism, but it is not in any direct sense citizenship as previously understood.

If volunteer tourism initiatives are where people look to act in relation to development, then that is in an important sense a restricted form of citizenship, as it bypasses the authority of the state, the latter having the potential to act as the political expression, democratically, of its citizens. The appeal of volunteer tourism to individuals as moral agents is worthy, but unmediated through an agonistic public sphere the parameters of this moral agency are extremely constrained. Criticising the claims made for global civil society in this vein, Chandler refers to a 'blurring [of] the distinction between the citizen, with rights of formal democratic accountability' on the one hand and the 'merely moral claims of the non-citizen' on the other (2004, p. 194). Chandler is not criticizing moral approaches *per se* here, but making a case for a political, or politically engaged, morality.

As Standish points out, global citizenship rhetorically eschews nationally based political channels (sovereign governments, unions etc) (Standish, 2012). It presents itself as having no axe to grind beyond that of the globe and the people living on it. It does not require political judgement, but instead emphasises awareness, responsibility and caring. This personalisation of the issues and the attendant encouraging of such private virtues is characteristic of volunteer tourism. Indeed, care, awareness and responsibility are the key aspects of its claim to facilitate social agency, or 'making a difference' (Butcher, 2015).

It is worth commenting on an irony of global citizenship evident in volunteer tourism: that a promotion of the value of *localism* (through projects that are invariably local and small scale and that often make a virtue of this) seems to sit easily with an advocacy of *global* citizenship. Here, what the local and the global share is a deprioritisation of the nation and national politics. The nation is on the one hand seen as too big to reflect local concerns, but too small to reflect global imperatives. Yet this deprioritisation of the nation state also reflects the deprioritisation of the demos, democracy and politics itself. Global citizenship is a citizenship cut adrift from the democratic contestation of society.

Hannah Arendt's political philosophy sheds light on the limitations of global citizenship and on volunteer tourism's capacity to contribute to political enlightenment. She argued that the full realisation of human freedom requires the development of a public realm. Such a realm historically represents the extension of human freedom beyond the private sphere of the family and intimate life. It brings together the diversity of private experience and interest into an agonistic public space. Whilst 'everyone sees and hears from a different position' (Arendt, 1958, p. 57) through private experience, this public space is the basis for striving for a 'word in common' (Arendt, 1958, p. 58). In modern societies the public realm is defined by the citizenship of a state. Arendt's republican citizen is an active part of the political determinations of states, states being the principal institutions of power and authority.

A citizenship outside the state is therefore a limited citizenship, unable to truly strive for a common world. Outside an agonistic public sphere enabling the political contestation of ideas and power, private virtues are projected onto human problems unmediated by a political framing. Without the potential for politics to transcend or mediate differences, private experiences (by their very nature differential and varied) dominate.

For Arendt, 'freedom as a demonstrable fact and politics coincide and are related to each other like two sides of the same matter' (Arendt, 2000, p. 442). This is apposite with regard to volunteer tourism. Freedom to act without politics is an attenuated freedom. Despite the widespread rhetoric on the theme of developing one's ethical identity through travel and experience, individuality is in fact limited by the emphasis on self-development.

Global citizenship involves a shift away from the potentially political citizenship seen as vital by Arendt to a moral one, where morality is set apart from the contestation of ideologies and power. It is, for Arendt through the process of politics that different societies and interests can try to achieve a 'world in common', itself a truly moral goal (Arendt, 1958). Global citizenship circumvents political power in the name of 'the globe', replacing it with pre-political virtues such as respect, care and responsibility, exercised by individuals through the market and the non-governmental organisation sector.

Arendt's advocacy of republican citizenship is apposite, too, in relation to the notion of volunteer tourism as neo-colonial. The neo-colonial critique holds that acts of benevolence that are not sensitized to the legacy of colonialism reproduce an active caring western subject and a passive grateful southern one (Mostafanezhad, 2014; Vrasti, 2012). Anti-colonial movements of the past demanded national sovereignty, and the national citizenship accompanying this represented freedom from colonial subjection. It was celebrated as such across former colonies on independence. Yet global citizenship bypasses not only the polities of the voluntourists, but also those of the nations being visited. In fact there is sometimes an implied or explicit criticism of nations in the global south accompanying volunteer tourism: their capacity to run orphanages, deal with social problems and educate their people.

A restatement of the importance of national sovereignty and national citizenship, in the countries of the global south as well as those in the global north, challenges the undermining of the sovereignty and agency rightly noted in the critiques of Vrasti (2012) and Mostafanezhad (2014). It re-emphasises sovereignty, both of the individual citizen and of the nation, and the potential for people to be agents of their own destiny through politics. It de-emphasises the political import of volunteers' acts of care, acts that are detached from the formal and substantial institutions of power and politics that shape the realities of both the tourist and their hosts.

A more cosmopolitanism tourist?

Underlying the advocacy of global citizenship is a sense that national citizenship is limited and parochial, and that global citizenship has the potential to overcome these limits and be part of a more cosmopolitan outlook. This is simplistic. Thomas Paine famously said in *The rights of man* (1781/2000): 'The world is my country, all mankind are my brethren, and to do good is my religion', yet spent his adult life agitating for republican citizenship in the USA, France and the UK, precisely so free citizens could shape their destiny and 'do good'. As Parekh argues, global politics may be better approached through a citizenship defined by a focus on political power and the institutions that wield it (2003). In other words, although it may seem counter-intuitive, a *global* orientation may be better served by a reinvigoration of *national* citizenship. Global citizenship inspired volunteer tourism does not facilitate privileged access to global understanding.

Also political campaigns and engagement have often taken the globe as their remit – this was the case well before global citizenship. For example, domestic issues in the nineteenth century such as bread prices were both national and global, influenced by grain imports and the duties levied. The political debates around these were shaped by this truism. Colonialism and imperialism were justified with reference to the globe, as was the opposition to them.

On the Left of politics, there is a tradition of internationalism, borne out of the belief that workers have no country and are united, globally, by their position in relation to their employers and capitalism, and their potential to advance society (Wilde, 2013). Capitalism too has been justified with reference to its capacity to develop the globe and safeguard freedom around the world. National citizenship has never precluded global concern. The argument that today society faces intertwined and complex so-called wicked problems that are best addressed by global citizenship is difficult to sustain – were not the wars, famines and epidemics of the past as complex and severe as those today?

The principal difference between political movements of the past and those influenced by global citizenship today is that the former addressed these global issues through the contestation of politics, not through ethical imperatives to care or act responsibly alone.

It is precisely the crisis of citizenship itself that has led to rise of global citizenship (Standish, 2012). The citizen is no longer linked to society through the institutions of politics, as in the past. The public sphere seems empty and uninspiring to many (Swyngedouw, 2011). Ideologies that facilitated political judgement are exhausted and new ideas and movements that might serve that function have not emerged (Chouliaraki, 2013; Furedi, 2013). Global citizenship bypasses the public sphere and connects private feelings and qualities such as care, empathy and awareness, with the global issues of the day (Popke, 2006). Hence these issues are reinterpreted as issues of personal ethics rather than as issues of political contestation. The trend towards anti-politics that many argue defines the period we live in is only reinforced by this trend (Clarke, 2015).

Consider, for example, the category of solidarity, strongly implicit in volunteering overseas (Lewis, 2006). Critical citizens in the past may have shown solidarity (albeit perhaps all too rarely) with their peers in poorer countries on a political basis through a common identification of social class or interest, or through a recognition that the oppression of a people elsewhere strengthened the hand of rulers at home (Wilde, 2013). By contrast, solidarity today, through lifestyle political practices such as Fairtrade, is an adjunct to consumption, a fleeting moment of charity towards other individuals (Chouliaraki, 2013). Volunteer tourism arguably involves greater dedication, but the association of social action with leisure travel betrays a similar lack of ambition and commitment.

Volunteer tourism's pragmatic politics of the possible

An appealing feature of global citizenship is that it facilitates social action where political action may seem untenable – it is 'doable'. For schools, global citizenship focuses on recycling, responsible shopping and Fairtrade. Relevance to daily life, practicality and pragmatism are attractive features. As a global citizen, one can always 'do your bit' for the planet or for others.

Volunteer tourism fits this pattern. To buy the right holiday, to help build a school, to hug a distressed child – to *do* something – replaces the more distant, seemingly untenable and for some undesirable collective and political project of shaping transformative development and promoting economic growth.

Personal responsibility in the face of major global threats is a common theme in global development education – 'what would / will you do' is implicit (Bourn, 2014; Standish, 2012). Standish outlines the tendency in global citizenship education to personalise and make development issues immediately relevant to the life and lifestyle of the individual (2012). Clearly this approach is attractive, and the implicit call to take things on personally is laudable. To be able to act and witness or at least visualise the outcome of one's actions can be inspiring and many volunteer tourists find it so (Bourn, 2014).

The problem is that what appears possible, or 'doable', in antipolitical times is very limited. It places agency squarely in the context of one's own biography, one's own lifestyle, rather than in the context of the individual's capacity to challenge entrenched political ideas and institutions. Whilst taking personal responsibility is a progressive impulse, in the advocacy of global citizenship it is also *private* responsibility – responsibility posed in the context of one's lifestyle, consumption decisions and emotions cut adrift from a political framing.

Volunteer tourism appeals to the impulse to act in pursuit of a better world, a commendable impulse at all times. Where that impulse has few inspiring outlets through politics, it manifests itself increasingly through lifestyle and the rhythms of everyday life, very much in keeping with Giddens' notion of 'Lifestyle Politics' (Giddens, 1994). The breadth of innovative ways to make a difference through lifestyle – shopping, telethons, wrist bands, volunteer tourism – correspond to the lack of ambition and vision with regard to what that difference might actually be.

For Arendt, a 'world in common' can only be constructed out of political contestation in the public sphere (Arendt, 1958). There is no global public sphere. Newspapers, political parties and trade unions are nationally based, and citizens vote in national elections for national parliaments. 'What is possible' needs to be challenged and expanded. It is therefore necessary to restate the importance of citizenship itself, rather than global citizenship, in order to challenge the privatisation and depoliticisation of development issues that volunteer tourism is indicative of.

Volunteer tourism: outsourcing citizenship to 'the globe'?

Citizenship has historically referred to the relationship between the nation state and individual citizen of that state. The shape of that relationship has changed over time. However, citizenship as a national phenomenon has never precluded global or international political concerns. *National* citizenship has been the focus for *global* political issues.

Citizenship as a normative category assumes that the individual citizen is involved in the politics of their nation state. In recent decades the institutions through which this political citizenship functioned feel like empty shells, and formal engagement tends to be low. Behind this lies a pervasive crisis of meaning and a lack of vision as to what the future could or should look like (Furedi, 2013; Laidi, 1998). There is a no clear moral framework on offer through citizenship (Furedi, 2013). Moral and ethical strategies are unlikely to be linked to national politics, and are far more likely to be associated with disparate campaigns and lifestyles (Giddens, 1994; Kim, 2012).

The crisis of meaning at the heart of politics has led elites to look elsewhere for some sort of moral purpose or justification (Chandler, 2004; Laidi, 1998). As a result, they have been keen to endorse global citizenship as a focus in education and in general (Standish, 2012). There is a sense in which the process of producing citizens is being outsourced from the nation to the globe, from the institutions of the state to companies and non-governmental organisations.

The growth of volunteer tourism is a good example of this outsourcing of citizenship. The global South has become a stage for the working out of what it is to be an ethical person (Chouliaraki, 2013). A number of writers and commentators on volunteer tourism have noted the way this outsourcing of citizenship functions in terms of a new political elite. Diprose points out that through international volunteering the global south acts as a 'training ground' for a new liberal elite for business and politics (2012, p. 190). For Pearce and Coghlan, volunteer tourism enriches the sending society by developing a 'pool of personnel with experiences and an embodied awareness of global issues' (2008, p. 132). For some, the gap year project is a part of building a portfolio of ethical experiences that shape the individual for a career in business or politics (Heath, 2007). Elsewhere it may be an 'immersion' experience to develop empathy with people who may be affected by decisions the global citizen makes in his or her career.

Material development benefits to the global South are minimal (Palacios, 2010). This suggests that the attempt to make a difference through ethical travel is very much a process driven by the crisis of politics in the west and the consequent search for meaning away from the institutions of politics. In the past if there was no development, then a development project would be said to have failed. Now it is legitimate to see the value in terms of the transformation of the volunteer and their personal journey towards global citizenship.

However, it has been argued that the contribution to development cannot be measured simply in terms of the projects themselves (Simpson, 2004). Rather, the projects play a role in developing people who will, in the course of their careers and lives, act ethically in favour of the poor and the oppressed. Thus, the experience of volunteering becomes 'an ongoing process which extends far beyond the actual tourist visit' (Wearing, 2001, p. 3).

For example, Chris Brown of Teaching Projects Abroad makes the case that a lack of experience of societies in the global south on the part of the bankers and businessmen of tomorrow contributes to exploitative relationships:

> How much better it might have been if all the people who are middle and high management of Shell had spent some time in West Africa [...] how differently they would have treated the Ibo people in Nigeria? [sic]. (cited in Simpson, 2004, p. 190)

Jonathan Cassidy of Quest Overseas concurs, arguing that if influential business people could only:

> look back for a split second to that month they spent working with people on the ground playing football with them or whatever' then they would act more ethically in their business lives. (cited in Simpson, 2004, p. 191).

Such sentiments are typical: through individual experience we can develop, decision by decision, a more ethical world with less suffering, more fairness and greater opportunity.

There is a narcissism to this outsourced search for moral meaning. It leads away from addressing the pressing material needs of others in the context of *their* lives and towards addressing the crisis of political identity in the West (Chouliaraki, 2013). The claim that volunteers' ethical careers post-trip can lead to change is false. It simply repeats the cycle of lifestyle-oriented individual strategies that view the individual not as a citizen within a polis, but as an employee or consumer within the global market. It only promotes further lifestyle political initiative, and fails to contribute to debate on development.

Conclusion

This aim of this paper is not to argue against volunteer tourism *per se*, but to criticise the tendency to see it as part of a new, progressive focus for citizenship: global citizenship. Travel can certainly broaden the mind and the impulse to help is of course progressive. However, global citizenship through volunteer tourism is questionable as a normative goal. It focuses the desire to act away from political citizenship, which in a world of nation states inevitably has a strong national dimension. In its place, the engaged citizen is encouraged to act through the rhythms of his/her life – lifestyle and consumption – via non-governmental organisations and private companies. Hence even leisure – holidays in this case – is associated with social agency through its contribution to global civil society and global citizenship.

The result of this is to elide the private virtue of care with the public question of development and to substitute the personal ambition to do good for the political question of social change. Care and the desire to 'make a difference' are laudable human qualities. Through the narrative of global citizenship, they are substituted for reflection upon, and political contestation of, the reasons for the poverty volunteer tourists are often reacting to.

A restatement of the importance of republican citizenship is a useful way to look at the limitations of global citizen-oriented volunteer tourism. Republican citizenship redirects agency from unaccountable companies and non-governmental organisations to the principal institutions of sovereign democracy and political power: state governments. Republican citizenship also assumes a respect for the citizens of other societies as sovereign political actors within a polity, and not recipients of lifestyle largesse through the market or non-governmental organisations.

Disclosure statement

No potential conflict of interest was reported by the author.

References

Apale, A., & Stam, V. (Eds.). (2011). *Generation NGO*. Toronto: Between the Lines.
Arendt, H. (1958). *The human condition*. Chicago, IL: University of Chicago Press.
Arendt, H. (2000). What is freedom? In *The portable Hannah Arendt* (pp. 438–461). London: Penguin Books.
Association for Tourism and Leisure Education/Tourism Research and Marketing. (2008). *Volunteer tourism: A global analysis*. Arnhem: Association for Tourism and Leisure Education.
Bianchi, R., & Stephenson, M. (2014). *Tourism and citizenship: Rights, freedoms and responsibilities in the global order*. London: Routledge.
Bourn, D. (2014). *The theory and practice of global learning*. (DERC Research Paper No. 11). London: IOE.
Broad, S. (2003). Living the Thai life – a case study of volunteer tourism at the Gibbon Rehabilitation project, Thailand. *Tourism Recreation Research*, 28(3), 63–72.
Butcher, J. (2015). Ethical tourism and development: The personal and the political. *Tourism Recreation Research*, 40(1), 71–80.
Butcher, J., & Smith, P. (2010). 'Making a difference': Volunteer tourism and development. *Tourism Recreation Research*, 35(1), 27–36.
Butcher, J., & Smith, P. (2015). *Volunteer tourism: The lifestyle politics of international development*. London: Routledge.
Chandler, D. C. (2004). *Constructing global civil society: Morality and power in international relations*. Basingstoke: Palgrave Macmillan.
Chen, S. (2002). Transitions to civic maturity: Gap year volunteering and opportunities for higher education. In W. Paxton & V. Nash (Eds.), *Any volunteers for the good society?* (pp. 67–77). London: Institute for Public Policy Research.
Chouliaraki, L. (2013). *The ironic spectator: Solidarity in the age of post humanitarianism*. London: Polity.
Clarke, N. (2015). Geographies of politics and anti-politics. *Geoforum*, 62, 190–192.
Crabtree, R. (2008). Theoretical foundations for international service-learning. *Michigan Journal of Community Service Learning*, 15(1), 18–36.
Delanty, G. (2000). *Citizenship in a global age*. London: Oxford University Press.
Devereux, P. (2008). International volunteering for development and sustainability: Outdated paternalism or a radical response to globalization? *Development in Practice*, 18(3), 357–370.
Dill, J. S. (2013). *The longings and limits of global citizenship*. London: Routledge.
Diprose, K. (2012). Critical distance: Doing development education through international volunteering. *Area*, 44(2), 186–192.
Dower, N. (2003). *An introduction to global citizenship*. Edinburgh: Edinburgh University Press.
Faulks, K. (2000). *Citizenship*. London: Routledge.
Furedi, F. (2013). *Authority: A sociological history*. Cambridge: Cambridge University Press.
Giddens, A. (1994). *Beyond left and right: The future of radical politics*. Oxford: Polity.
Heater, D. (2004). *A brief history of citizenship*. New York: New York University Press.
Heath, S. (2007). Widening the gap: Pre-university gap years and the 'economy of experience'. *British Journal of Sociology of Education*, 28(1), 89–103.
Howes, A. J. (2008). Learning in the contact zone: Revisiting neglected aspects of development through an analysis of volunteer placements in Indonesia. *Compare*, 38(1), 23–38.
Jacoby, R. (1999). *The End of utopia: Politics and culture in an age of apathy*. New York, NY: Basic Books.
Jakubiak, C. (2012). 'English for the globe': Discourses in/of English language voluntourism. *International Journal of Qualitative Studies in Education*, 25(4), 435–451.
Jones, A. (2011). Theorising international youth volunteering: Training for global (corporate) work? *Transactions of the Institute of British Geographers*, 36, 530–544.
Kim, Y. M. (2012). The shifting sands of citizenship toward a model of the citizenry in life politics. *Annals of American Academy of Political and Social Science*, 644(1), 147–158.
Krippendorf, J. (1987). *The holidaymakers: Understanding the impact of leisure and travel*. London: Butterworth-Heinemann.
Laidi, Z. (1998). *A world without meaning: The crisis of meaning in international politics*. London: Routledge.
Lewis, D. (2006). Globalisation and international service: A development perspective. *Voluntary Action*, 7(2), 13–26.
Lyons, K., Hanley, J., Wearing, S., & Neil, J. (2012). Gap year volunteer tourism: Myths of global citizenship? *Annals of Tourism Research*, 39(1), 361–378.
Lyons, K., & Wearing, S. (2011). Gap year travel alternatives: Gen-Y, volunteer tourism and global citizenship. In K. A. Smith, I. Yeoman, C. Hsu, & S. Watson (Eds.), *Tourism and demography* (pp. 101–116). London: Goodfellow Publishers.

McGehee, N. (2002). Alternative tourism and social movements. *Annals of Tourism Research, 29*, 124–143.

McGehee, N., & Santos, C. (2005). Social change, discourse, and volunteer tourism. *Annals of Tourism Research, 32*(3), 760–776.

McGehee, N. G. (2012). Oppression, emancipation, and volunteer tourism: Research propositions. *Annals of Tourism Research, 39*(1), 84–107.

Mostafanezhad, M. (2013). The politics of aesthetics in volunteer tourism. *Annals of Tourism Research, 43*, 150–169.

Mostafanezhad, M. (2014). *Volunteer tourism: Popular humanitarianism in neoliberal times*. Farnham: Ashgate.

Mowforth, M., & Munt, I. (2015). *Tourism and sustainability: New tourism in the third world*. London: Routledge.

Novelli, M. (2004). *Niche tourism: Contemporary issues, trends and cases*. London: Routledge.

Oxfam. (in press). *Global citizenship*. Retrieved October 17, 2014, from http://www.oxfam.org.uk/education/global-citizenship

Paine, T. (1781/2000). *The rights of man*. New York, NY: Dover Publications.

Palacios, C. (2010). Volunteer tourism, development, and education in a postcolonial world: Conceiving global connections beyond aid. *Journal of Sustainable Tourism, 18*(7), 861–878.

Parekh, B. (2003). Cosmopolitanism and global citizenship. *Review of International Studies, 29*, 3–17.

Paxton, W., & Nash, V. (2002). *Any volunteers for the good society*. London: Institute for Public Policy Research.

Pearce, P. L., & Coghlan, A. (2008). The dynamics behind volunteer tourism. In K. D. Lyons & S. Wearing (Eds.), *Journeys of discovery in volunteer tourism: International case study perspectives* (pp. 130–143). London: CABI.

Peters, M., Britton, A., & Blee, H. (2007). *Global citizenship education*. London: Sense Publishers.

Phi, G., Dredge, D., & Whitford, M. (2013, April 13–16). *Fostering global citizenship through the microfinance-tourism nexus*. Paper presented at Tourism education for global citizenship: Educating for lives of consequence conference. Oxford Brookes University.

Popke, J. (2006). Geography and ethics: Everyday mediations through care and consumption. *Progress in Human Geography, 30*(4), 504–512.

Raymond, E. (2008). Make a difference! The role of sending organizations in volunteer tourism. In K. D. Lyons & S. Wearing (Eds.), *Journeys of discovery in volunteer tourism: International case study perspectives* (pp. 48–62). London: CABI.

Rhoads, R. A., & Szelenyi, K. (2011). *Global citizenship and the university: Advancing social life and relations in an interdependent world*. Stanford, CA: Stanford University Press.

Simpson, K. (2004). *Broad horizons: Geographies and pedagogies of the gap year* (Unpublished PhD thesis). University of Newcastle, Callaghan.

Simpson, K. (2005). Dropping out or signing up? The professionalisation of youth travel. *Antipode, 37*, 447–469.

Sin, H. L. (2009). Volunteer tourism: 'Involve me and I will learn'? *Annals of Tourism Research, 36*(3), 480–501.

Sin, H. L. (2010). Who are we responsible to? Locals' tales of volunteer tourism. *Geoforum, 41*(6), 983–992.

Smith, K., & Holmes, K. (2009). Researching volunteers in tourism: Going beyond. *Annals of Leisure Research, 12*(3/4), 403–420.

Söderman, N., & Snead, S. (2008). Opening the gap: The motivation of gap year travellers to volunteer in Latin America. In K. D. Lyons & S. Wearing (Eds.), *Journeys of discovery in volunteer tourism: International case study perspectives* (pp. 118–129). London: CABI.

Standish, A. (2008). *Global perspectives in the geography curriculum: Reviewing the moral case for geography*. London: Routledge.

Standish, A. (2012). *The false promise of global learning: Why education needs boundaries*. London: Continuum.

Swyngedouw, E. (2011). Interrogating post-democracy: Reclaiming egalitarian political spaces. *Political Geography, 30*, 370–380.

Tomazos, K., & Cooper W. (2012). Volunteer tourism: At the crossroads of commercialisation and service? *Current Issues in Tourism, 15*(2), 405–423.

Vrasti, W. (2012). *Volunteer tourism in the global south: Giving back in neoliberal times*. London: Routledge.

Wearing, S. (2001). *Volunteer tourism: Experiences that make a difference*. Wallingford, CT: CABI.

Wearing, S. (2003). Volunteer tourism. *Tourism Recreation Research, 28*(3), 3–4.

Wearing, S., & Darcey, S. (2011). Inclusion of the "othered" in tourism. *Cosmopolitan Civil Societies Journal, 3*(2), 18–34.

Wearing, S., & Grabowski, S. (2011). Volunteer tourism and intercultural exchange: Exploring the 'other' in this experience. In A. M. Benson (Ed.), *Volunteer tourism: Theoretical frameworks and practical application* (pp. 193–210). London: Routledge.

Wearing, S., McDonald, M., & Ponting, T. (2005). Building a decommodified research paradigm. *Journal of Sustainable Tourism, 13*(5), 424–439.

Wearing, S., & McGehee, G. (2013). *International volunteer tourism: Integrating travellers and communities*. Wallingford, CT: CABI.

Wilde, L. (2013). *Global solidarity*. Edinburgh: Edinburgh University Press.

Yanapuma Foundation. (in press). *Global citizenship – concepts and practice*. Retrieved December 16, 2014, from http://www.yanapuma.org/en/resintoverview.php

A rite of passage? Exploring youth transformation and global citizenry in the study abroad experience

Simone Grabowski, Stephen Wearing, Kevin Lyons, Michael Tarrant and Adam Landon

ABSTRACT
Travel, long recognised as a *rite of passage*, is often also touted as a transformative experience which facilitates cross-cultural understanding, fosters an embrace of diversity and promotes global awareness. This process is aligned with youth development and has a rich history in the tourism literature. The importance of transformational travel, however, has now spread to programmes across the higher education landscape, with the recognition that travel has the potential to nurture a global citizenry. Additionally, for many young people, the motivation for studying abroad is to assist in the transition to adulthood. In this way, educational travel is similar to an 'overseas experience' or a 'gap year'. It is often taken at an important time of transition in emerging adulthood, for example, from school to work. We argue that this period of identity formation for youth can be likened to a rite of passage much like the Grand Tour of the seventeenth and eighteenth centuries was for young European men and women. Our paper examines the role of the study abroad experience in promoting youth transformation and global citizenry.

Introduction

Temporary transnational moves have been increasingly seen as important in the transition to adulthood of young people in Western nations (Cairns, 2008; Thomson & Taylor, 2005; Yoon, 2014). It has been suggested that short-term youth expeditions, characteristic of study abroad experiences, closely mimic 'rites of passage' for young individuals who have the opportunity to learn about themselves (Bagnoli, 2009; Beames, 2004; King, 2011; Starr-Glass, 2016). This idea has been linked in recent years to the importance of study abroad experiences contributing to the development of a global citizenry (Tarrant, 2010). Our paper provides an initial exploration of the relationship between self and identity, global citizenship and educational travel for the 18–30-year-old demographic and links it back to the ideas found in the Grand Tour of the seventeenth and eighteenth centuries and van Gennep's (1960) conceptualisation of the 'rites of passage' in an attempt to provide frameworks for future research.

With the rapid increase of youth participation in travel, we examine historical motives and hypothesised outcomes (i.e. global citizenship (Adamson & Ruffin, 2013; Cameron, 2013)) associated with international travel, and their similarities and differences with modern study abroad experiences. Specifically, we explore the extent to which modern educational travel can be linked to the early ideas of the Grand Tour and may have become the new rite of passage for youth in the twenty-second century. Youth competence in this society differs, however, status and identity may still be obtained through travel as a rite of passage. Our purpose here is to examine the links to provide a framework that, as Stoner et al. (2014) suggest, can be used to undertake research to provide evidence that the assumed outcomes of study abroad experiences, including changing self-concepts, are actually being achieved. Exploring the identity formation process in the context of global educational travel (GET) can yield recommendations for pedagogy that maximise student growth and learning, while identifying new questions and concerns for future research and practice.

In this paper GET refers to all overseas travel undertaken by enrolled undergraduate students for the purpose of study, and the global competencies (values, identities and knowledge) related to global citizenship, security and international relations, student engagement and work adaptability that stem from it. Although there has been increased participation in GET in recent years, the duration of students' overseas stays has fallen dramatically (Dwyer, 2004; Institute of International

Education, 2015). Consequently, traditional learning outcomes associated with study abroad experiences (e.g. language acquisition) developed over longer periods of time have, in many ways, become secondary in importance to emerging metrics of students' global competencies, including aspects of the self. However, greater clarity is needed to understand the role of GET in fostering higher order learning outcomes including students' post-cosmopolitan self-concepts (Landon, Tarrant, Rubin, & Stoner, 2017; Sutton & Rubin, 2004). We see that an area in which such GET programmes perhaps have the greatest potential impact, particularly with respect to the mission statements of universities and colleges, is in nurturing a global citizenry (Dolby, 2007). We place our exploration of GET as a 'rite of passage' within this context.

Global citizenship

Tarrant et al. (2014) suggest that citizenship refers to a national identity with special rights and duties prescribed by the respective government; as such a global citizenry cannot be similarly characterised since there is no global government and/or few enduring international laws (Noddings, 2005). Though competing definitions of global citizenship abound (Davies, Evans, & Reid, 2005) there are three key dimensions (or obligations) which are now generally accepted in the literature: social responsibility, global awareness and civic engagement (Morais & Ogden, 2011). Schattle (2009, p. 12) proposes that global citizenship 'entails being aware of responsibilities beyond one's immediate communities and making decisions to change habits and behavior patterns accordingly', while (Galston, 2001, p. 217) acknowledges that 'it is reasonably clear that good citizens are made, not born. The question is how, by whom, and to what end?'

Cultural differences are an important part of global citizenship education and policy which seeks to promote the celebration of cultural differences. This is an ideology that has underpinned Australian government multiculturalism policies (Ozdowski, 2012) for the past two decades. Community and political leaders who espouse this ideal argue that it is central to the development of tolerance and acceptance that underpins community safety, and by extension, national security and enables Australians to engage fully in a global market economy (Modood, 2007; Ozdowski, 2012). However, both historic and recent national incidents of violence associated with race and cultural relations that have attracted national media attention, call into question whether such an ideal is little more than empty rhetoric. Acceptance of cultural diversity, a central tenet to global citizenry (Rubin, Landon, Tarrant, Stoner, & Mintz, 2016), clearly does not occur through a process of social osmosis but the nature of the active learning process that lead people to become global citizens is not well understood.

The nature, values, attributes and efficacy of global forms of identification and community are much debated and incorporate notions of cosmopolitan (Nussbaum, 1996) or transnational citizenship (Habermas, 2001) as well as global citizenship (Dower, 2000). The challenges posed by climate change, international population flow, cross-border exchanges and equitable distribution of international resources are undeniably global in scope and impact and challenge the very notion of a nationally bounded citizenry (Kofman, 2005). Indeed, it has been argued that in the twenty-first century the nation is no longer the exclusive framework for social, cultural and political identification (Banks, 2004, 2009). According to these discourses, citizenship seems to be shifting scale, moving away from national affiliations and towards global forms of belonging, responsibility and political action.

International travel is potentially an important facilitator of global citizenship as it creates exchange between cultures and through the transfer of populations across states, regions and cultures accounting for over one-twelfth of world trade constituting by far the largest movement of people across borders. The impact of international tourism and global mobility is far-reaching across economic, social and cultural life. International and domestic tourism account for 10% of global employment and global GDP with 1186 million international passenger arrivals each year, predicted to reach 1.8 billion by 2030 (UNWTO, 2016). GET accounts for a small percentage of this but is growing rapidly.

Global educational travel

For many years now educators have been faced with a challenge to produce more socially and globally engaged students (Liang, Caton, & Hill, 2015). They have turned to modes of learning which span beyond the classroom, and as such the higher education landscape has incorporated experiential learning in its curriculum which includes field trips and internships. The GET (also known as study abroad or student exchange) phenomenon has formed part of this shift. Here it is noted that experiential learning is implicit within virtually all forms of travel (i.e. the very nature of boarding a plane and entering a foreign destination through customs is an experience in itself) and this inferred or assumed outcome has emerged over the past two decades in the tourism literature on the 'experience economy'

(Larsen, 2007; Mehmetoglu & Engen, 2011; Pine & Gilmore, 1999). However, it is experiential learning that goes beyond passive acquisition that is transformative. The structuring of experiential learning has been central to both formal experiential education programmes and the positive leisure movement (Larson, 2000). One of the shared features of such programmes has been the use of reflective learning (Moon, 2004) to bring about transformational outcomes.

Interest and growth in GET, and its link to global citizenship, is evidenced by developments at a number of social, political and institutional levels. In the USA, the bi-partisan Lincoln Commission (2005, p. 3) recognised that 'what nations don't know can hurt them … the stakes involved in study abroad are that simple. For their own future and that of the nation, it is essential that college graduates today become globally competent'. The Australian Government's *National strategy for international education 2025* (2016, p. 7) recognises that an international education 'offers opportunities to build enhanced bilateral and multilateral relationships, which increase cultural awareness and social engagement'. A primary mechanism for implementation of policy on global education has been investment in outbound mobility programmes for university students. In the USA, the Senator Paul Simon Study Abroad Program Act (S. 3390) has established a target of one million US students studying abroad for credit, up from the current ~300,000. In Australia, initiatives such as the New Colombo Plan are expanding from a pilot phase to full implementation in the next few years with A$160 M committed to enable more than 10,000 Australian students to participate in outbound mobility programmes by 2019.

Since the first major data collection on Australian student outbound mobility was undertaken in 2005, the percentage of undergraduates who participate in these programmes during their degree programme has more than doubled, from 5% (Olsen, 2008) to 13.1% in 2012 (Australian Government, 2014). In the USA, 2013/2014 data showed 14.8% of undergraduate students had studied abroad for academic credit, an increase of 5.2% over the previous year, with US undergraduate participation in study abroad more than tripling over the prior two decades (Institute of International Education, 2007, 2015).

Travel, as a part of a global education, has both nation state and individual dimensions. For example, the USA in its desire to ensure that it remains competitive in an increasingly global marketplace while responding to global issues, conflicting resource utilisation, economic demands and national security underlies global competence and national needs (Lewin, 2009). On a personal level, global citizenship has been defined as a 'meritorious viewpoint that suggests that global forms of belonging, responsibility, and political action counter the intolerance and ignorance that more provincial and parochial forms of citizenship encourage' (Lyons, Hanley, Wearing, & Neil, 2012, p. 361). In bringing these interests together, the young global citizen of today might be a captain of industry tomorrow bringing positive change on a global scale (Lyons, 2015). Empirical research, for instance, has demonstrated that a self-concept that incorporates aspects of the global community is an antecedent to pro-social values, attitudes and behaviours (Reysen & Katzarska-Miller, 2013). Therefore, transnational identities developed through GET may foster individual, social and corporate responsibility as students assume positions of power.

We emphasise that using global travel as a conduit for fostering a global citizenry is a not a new idea. For example, Diprose (2012) has argued that students interested in development studies would be best to enrol in courses which are formally structured to provide an overseas educational opportunity which does not perpetuate the stereotypes of the global South such that many volunteer tourism programmes have been criticised for when promoting the benefits of a transformative experience (Butcher, 2011; Guttentag, 2009; McBride, Brav, Menon, & Sherraden, 2006; Simpson, 2004). In this way, these components of the course can facilitate the enactment of global citizenship, mutual exchange, reciprocity and ethical concerns for social justice (Jorgenson, 2010). A structured course has been advocated to ensure academic rigour, learning outcomes and goals are achieved (Dall'Alba & Sidhu, 2015; Stoner et al., 2014).

GET has provided academic institutions with a platform to potentially foster global citizenship and produce globally competent graduates (Stoner et al., 2014). Over 90% of Australian universities have strategic plans that support global mobility (AIM Overseas, 2011). While international service learning, is increasingly seen as a non-negotiable component of many undergraduate degrees and increasingly, students who undertake university-sponsored GET do so with the understanding that such experience is essential for one's education and future career (Daly, 2011; Forsey, Broomhall, & Davis, 2012). For example, GET has been found to provide beneficial learning outcomes including the improvement of student grades (Sutton & Rubin, 2004, 2010). Further, several studies have shown that employers value a study abroad experience (Crossman & Clarke, 2010; Curran, 2007; Trooboff, Vande Berg, & Rayman, 2008) but there is very little evidence to suggest that students' future careers have developed as a result of the experience itself (Franklin, 2010). Instead, it is possible that top students are more attracted to GET in the first place due to their above average grades or levels of intercultural competence.

There are many other benefits of GET for the student. The notion that an overseas experience can increase ones intercultural competence has been supported by many studies (Clarke, Flaherty, Wright, & McMillen, 2009; Daly, 2011; Starr-Glass, 2016; Twombly, Salisbury, Tumanut, & Klute, 2012; Williams, 2005). This includes the improvement of students intercultural communication skills (Williams, 2005), intercultural sensitivity (Clarke et al., 2009), global mindedness (Clarke et al., 2009; Hadis, 2005) and open mindedness (Hadis, 2005).

However, there are several studies which do not provide empirical support for many of the pre-mentioned outcomes of GET. Some hypothesise that this is due, in part, to heterogeneity in GET experience programming and length. Recent data on outbound mobility, for instance, reflect an increasing emphasis that both Australian and US universities are placing on short-term GET programmes, especially during the summer break. In 2012, participation by Australian students in short-term outbound mobility programmes represented 56% of all outbound mobility programmes. In the US, participation in short-term (less than 8 weeks) programmes increased from 51.4% of all programmes in 2004/2005 to 62.1% in 2013/2014 (Institute of International Education, 2007, 2015). This reflects the application of neo-liberal values to young people's travel, leisure and educational practices and reinforces the notion that leisure time itself is a context for some nation building (Lyons et al., 2012).

A final outcome of GET, which will be explored in the following section, is the change in identity and personal development to the student. Many studies have found that the experience goes further than fostering a greater understanding of the global, to fostering a greater understanding of self (Bandyopadhyay & Bandyopadhyay, 2015; Dolby, 2007; Hadis, 2005). Much like the grand tour, GET is an opportunity for self-exploration. The consequences of providing structured opportunities for self-exploration through GET, however, remain largely unknown.

Travelling for self and identity

Sussman (2000) argues that culture is an important reference for self-definition. Casmir (1984, p. 2) defines cultural identity as 'the image of the self and the culture intertwined in the individual's total conception of reality'. Culture's effect on behaviour is a well-documented phenomenon, however, its effect on the self is less widely understood (Marsella, 1985). The work that has been done linking culture and personality in the mid-twentieth century and then later by Geertz (1973), Hall (1976) and others clearly demonstrates that perception of self and identification differ across cultural boundaries. For example, Hall (1976, p. 226) argues that in cultures with strong family bonds such as Chinese, Japanese, Arab villagers and Spanish in North and South America, as a 'child moves into the larger and more real world of the adult … he does not, even under normal circumstances, establish an identity separate from that of his community'. This is very different to children brought up in Western cultures and construe the self independently (Markus & Kitayama, 1991). 'We draw a line around the individual and say this is our basic entity – the building block of all social relations and institutions' (Hall, 1976, p. 231). A tourist's engagement with the 'other' in travel redefines this notion. As support, Montuori and Fahim (2004) acknowledge the work of Adler (1977) on cross-cultural transitions and Hall's (1959, 1976, 1983) work on culture as being pivotal in understanding the self. Montuori and Fahim (2004, p. 256) suggest that 'an exchange with persons from other cultures can be used as a vehicle for changing oneself in a potentially desirable manner'. In fact, relationships formed in tourism have the potential to assist the tourist in constructing an interdependent self with the assistance of the 'other:' 'Others thus participate actively and continuously in the definition of the interdependent self' (Markus & Kitayama, 1991, p. 227).

Social identity theory posits that individuals construct identities based on their membership in social groups and interactions with important others (Tajfel & Turner, 1979). Ward (2004) notes that social identification leads to effective intercultural relations and in turn, affects psychological and socio-cultural adaptation. Relating this to GET, the process of social identification is when young students begin to compare themselves to their peers and others in the cross-cultural context and in this they find a space for development of self. Hibbert, Dickinson, and Curtin (2013) explored the effects of interpersonal relationships in travel on the perception of self. They found that development and maintenance of relationships affects identity and in-turn travel behaviour. This is a recent movement in the literature towards understanding that identity can play a large part in the types of travel experiences people choose, and the psychosocial outcomes that stem from those travel experiences.

Research drawing on social identity theory has demonstrated that social identification influences altruistic behaviour (De Cremer & van Vugt, 1999). If a global identity represents the broadest scale of identity formation, fostering a global identity may be an important mechanism for enhancing cooperation in global social dilemmas including climate change and biodiversity loss (Buchan et al., 2011). Students engaged in GET are then constructing identities based on the altruistic value of 'other': their worldview expands and resonates with a global citizenry.

If we pursue this idea of a search for identity and self-understanding we see it is only a modern phenomenon. Baumeister (1986) explains that individuality and adolescence was absent in pre-modern times before industrialisation and the division of labour occurred. Before this life was organised collectively much like what is currently seen in East Asian and Southern European countries. 'The family filled a vital role [and] placed children in adult roles' (Baumeister, 1986, p. 104). Many adult decisions were made for the young person including marriage partners and occupation (Baumeister, 1986). The young adolescent was seen to transform to adulthood once these two events had occurred. However, with industrialisation in the nineteenth century, more career options opened up and adolescents were not limited to farming. The adult identity then 'became something chosen by the adolescent rather than arranged by parental decision and influence within a context of limited options' (Baumeister, 1986, p. 105).

By the time an individual turns 18 most of his/her physiological and neural development is complete. In many Western countries this age signifies the transition from adolescence to adulthood as it coincides with key events like the end of secondary schooling and new legal responsibilities, such as voting and consuming alcohol. However at this age many young people have yet to assume other adult responsibilities; for example, completing tertiary studies, marriage and childbirth. The uncertainty of the scheduling of these events makes for a very confusing period for young people. Additionally, the length of this period is unknown with Côté and Bynner (2008, p. 253) arguing that 'the transition to adulthood is now taking longer on average than in the past, delayed until the mid-twenties to late-twenties for a significant proportion of youth cohorts in many developed societies'. This search for identity and self is seen as a transition to adulthood and has been termed young adulthood (Erikson, 1968, 1980), emerging adulthood (Arnett, 2000, 2002) and post-adolescence (Du Bois-Reymond, 1998).

Beck (1992) introduced the concept of 'choice biographies' to explain that young people faced increasing responsibilities due to the greater number of choices. And with this growing number of options young people began to postpone the assumption of adult roles to much later in life (Shulman, Feldman, Blatt, Cohen, & Mahler, 2005). Young adults have therefore become future oriented with a view that they can personally/individually alter their life courses. This notion feeds into that of Erikson's work on identity and the life cycle. Erikson (1968, pp. 243–244) explored the youth of his day and explained that youth were discontented and craved active locomotion. This is reiterated by Minh-ha (1994, p. 21) who suggests that the craving (that Erikson writes about) is a characteristic of every traveller and plays an important part in identity formation.

> Every voyage is the unfolding of a poetic. The departure, the cross-over, the fall, the wandering, the discovery, the return, the transformation. If travelling perpetuates a discontinuous state of being, it also satisfies, despite the existential difficulties it often entails, one's insatiable need for detours and displacements in postmodern culture. The complex experience of self and other (the all-other within me and without me) is bound to forms that belong but are subject neither to 'home', nor to 'abroad'.

Youth transformation, rites of passage and GET

With increasing numbers of youth now attending tertiary educational institutions GET is becoming a part of the transitional nature of youth in this period, from adolescence to adulthood. It is associated with the redefinition of self-identity, which for young people is often constituted by experiences of anxiety and the possibility for change, or 'fateful moments' (Giddens, 1991). Desforges (2000, p. 936) argues that many young people consider travel as a rite of passage that 'provides answers to questions that are raised about self-identity at fateful moments'. Therefore, long-term independent travel simultaneously considered as educational and character-building, and 'is imagined as providing for the accumulation of experience, which is used to renarrate and represent self-identity' (Desforges, 2000, p. 942). Similarly, Elsrud (2001) describes the travel experience as a process of narrating self-identity. Drawing on Giddens's (1991) conceptualisation of identity as a 'self-reflexive project', Elsrud (2001, p. 598) regards 'the traveller as narrator and the journey as narrative'. She states that independent travel, such as long-term global backpacking, is often presented as an adventurous lifestyle, and independent travellers are accredited with increased knowledge, a stronger sense of identity and social status (Elsrud, 2001, p. 597).

The importance of travel on the young individual's identity and development of self has been explored in the literature (Bagnoli, 2009; Riley, 1988; Shulman, Blatt, & Walsh, 2006; Wearing, 2002). In particular, Bagnoli (2009) and Frändberg (2014) explain that the travel experience may correspond with a period of transition for the young person. A conclusion drawn in a study of students returning from youth expeditions was that these expeditions 'could assist people to move through this transition more quickly' (Allison, Davis-Berman, & Berman, 2012, p. 498). King (2011) believes that the gap year is a point of transition to adulthood in the life of a young person as it is a time when a young person grows and matures. It is also taken at an important time of transition in emerging adulthood; from school to tertiary education or work, and taken in the early 20s 'before long-term commitments are made to partners, starting families or establishing careers' (Wilson, Fisher, & Moore, 2009, p. 4). The experience of travelling at life transitions, which Arnett (2000, 2002) argues is happening at emerging adulthood, may indeed be compared with the '"initiation rites" of traditional societies, which would often include a process of separation from the previous environment' (van Gennep, 1960 cited in Bagnoli, 2009, p. 326). We make this link to help explain how the young person then undergoes a transition (in a liminal phase) and is reincorporated back into the home society with a new status (van Gennep, 1960).

We suggest that van Gennep's (1960) conceptualisation of the rites of passage is a useful analogy to explore some of the aspects of student identity in GET and provides the structure for our discussion. According to van Gennep, the rite of passage is an individual's changed social membership and relationships over time (Tinto, 1988) where they negotiate their place in society. A rite of passage is characterised by three phases. Phase one, separation, sees the GET student preparing to exit the home country by limiting their interaction with their social groups. They are detached from their former self via some kind of ritual, for example, a leaving party. The second phase, transition (liminality) is where the student is between two states of being. Here interaction in the new society includes forming relationships with local and study abroad students as well as local people. Adoption of new behaviours and expansion of knowledge occurs in order to 'fit in'. The third phase is incorporation where the student becomes competent in the new society and has adopted a new status as a study abroad student. He/she is a member of a new group and upon return home, takes this status with him/her.

Therefore, GET is a period of identity formation for youth. The student will not return home with the same status. Frick's (1983, 1987) concept of the 'symbolic growth experience' can be used to explain some of these findings where the experience is said to assist in learning and growth of the individual. GET can be likened to a rite of passage much like the Grand Tour of the seventeenth and eighteenth centuries was for young European men and women (Reau, 2012). In the seventeenth, eighteenth and part of the nineteenth centuries, youth leisure travel was performed in the Grand Tour of Europe for the purpose of education and self-development by the affluent youth. This was an unstructured travel experience of up to three years which was initially undertaken by the British. Therefore the trip began in the UK and moved over to the European continent via France, Switzerland, Germany and Italy (Towner, 1985). Young travellers were often chaperoned, they rented apartments for months at a time in cities such as Milan, Florence and Venice and tutors were sought. Through this tour, travel was intended to increase one's worldliness and sophistication and 'to confer the traveller with full membership into the aristocratic power structure' (Weaver & Lawton, 2010, p. 52).

Reau (2012, p. 14) explains that for youth undertaking a Grand Tour, 'travel offered something that could not be learned at home: it allowed the traveller to reflect on himself and his own society, guided by the thought of his eventual return home'. Consequently, the increased mobility garnered by travel is essential to developing inner selves and transitioning to adulthood (Thomson & Taylor, 2005). We see the similarities with GET being the alignment of education and self-development. GET is not just a 'holiday' but an active experiential learning experience that engages in and with, a separation from 'home' and the opportunity to explore ones idea of self and identity in a new environment.

We acknowledge the limitations that this conceptualisation of a comparison between the Grand Tour and GET bring; where the focus of both concepts is young people travelling from the global north to the south (Mowforth & Munt, 2015). However, we see that they have significant overlaps and these overlaps offer insights into identity formation and social membership of the youth of their time. A key overlap is the lack of diversity in participation; the Grand Tour was predominantly undertaken by young white aristocrats and GET has been found to attract students from white backgrounds (Institute of International Education, 2015) and perpetuate social and class stereotypes (Gerhards & Hans, 2013; M'Balia, 2013). Adler (1985) has argued though that even in the British working class, their 'tramping' for work which occurred at a similar period in time to the Grand Tour, resembled a rite of passage to adulthood. Additionally, Towner's (1985) analysis showed that the class of the tourist changed in the Grand Tour's 300-year history to include a larger number of middle-class participants.

Recent studies have shown that the benefits of GET described in this paper transcend race, ethnicity, gender, class and ability (Ablaeva, 2012; Smith, Smith, Robbins, Eash, & Walker, 2013; Wick, 2011) and that financial and academic barriers can be overcome (Kasravi, 2009; Murray Brux & Fry, 2010). Consequently, the rite of passage can be experienced by all engaged in GET. This is important to consider in the changing climate of GET. For example, the demand for such programmes from students in regions such as East Asia, Central and Eastern Europe and Sub-Saharan Africa is growing in comparison to North America and Western Europe. Altbach, Reisberg, and Rumbley (2009) discussed this as a south–north phenomenon. Additionally, while the Grand Tour was predominately undertaken by young men, GET programmes (and those affiliated such as volunteer tourism) have a much higher female to male ratio (Salisbury, Paulsen, & Pascarella, 2010). Within these trends we will continue to see study abroad students engage with the 'other' in their personal quest for transformation resulting in greater global citizenry.

Starr-Glass (2016) argues that as an undergraduate student one is stuck between two socially constructed rites of passage; graduation from school and from college/university. At both stages a student is required to transition and adopt a new identity. We posit that when GET is added to a student's study plan there is an extra 'rite of passage' that he/she will encounter which can prove to be disruptive to identity formation yet also results in a transition to a more globally oriented, outward focussed identity. This rite of passage then can be more powerful than the other two in producing a global citizen.

Conclusion

This paper has explored some of the outcomes of the study abroad experience on global citizenry and development of self. We suggest that to liken GET to a rite of passage in terms of its impact on identity and transformation provides some insights into how it can be analysed. Each student will engage differently with their study abroad experience – some will take a more active role in student or wider community life. These experiences will affect their engagement with community and citizenry on return home. Regardless of these differences, we argue that the experience is an important component in the transition to adulthood as has also been described by Thomson and Taylor (2005) in their reference to cosmopolitanism and mobility. We posit that the sequence of life events in emerging adulthood that has been theorised by Arnett (2000) and others is missing a vital stage: a study abroad experience. What the students gain is some insight into understanding of their self as it engages in the components of the cross-cultural interactions of GET albeit with the limitations that also come with these experiences (cf. Mowforth & Munt, 2015).

We suggest that a 'rite of passage' has the potential to enhance and foster a transformative experience that leads to a shift in perspective, awareness and worldview, and that the short-term educational travel programmes offer and provide a learning site for students to experience, grapple with, reframe and reflect on issues global in nature (Wearing, Tarrant, Schweinsberg, Lyons, & Stoner, 2015). But to achieve this, GET programmes need to be grounded by a sound pedagogical framework which (a) ensures academic rigour, (b) establishes and measures resultant learning outcomes and (c) ascertains whether proposed goals are achieved (Landon et al., 2017; Stoner et al., 2014). It is this more structured form of travel that we see as central to a move to a more global citizenship for youth today with the potential to form part of a young person's rite of passage. This is opposed to the unstructured travel of the past (such as the Grand Tour) as we suggest in this age of neo-liberal dominance, travel has been captured by the market and is so heavily influenced towards consumption and self-gratification that some of the elements of a rite of passage have been subsumed.

Disclosure statement

No potential conflict of interest was reported by the authors.

ORCID

Simone Grabowski http://orcid.org/0000-0001-7904-1240
Stephen Wearing http://orcid.org/0000-0002-5158-059X

References

Ablaeva, Y. (2012). *Inclusion of students with disabilities in study abroad: Current practices and student perspectives* (Master's thesis). University of Oregon.

Adamson, J., & Ruffin, K. N. (2013). *American studies, ecocriticism, and citizenship: Thinking and acting in the local and global commons.* New York, NY: Routledge.

Adler, J. (1985). Youth on the road. *Annals of Tourism Research, 12*(3), 335–354.

Adler, P. S. (1977). Beyond cultural identity: Reflections on cultural and multicultural man. In R. Brislin (Ed.), *Culture learning: Concepts, application and research* (pp. 24–41). Honolulu: University of Hawaii Press.

AIM Overseas. (2011). *Outbound mobility best practice guide for Australian Universities*. Canberra: Australian Government Department of Industry, Innovation, Science, Research and Tertiary Education (DIISRTE) and Australian Education International (AEI).

Allison, P., Davis-Berman, J., & Berman, D. (2012). Changes in latitude, changes in attitude: Analysis of the effects of reverse culture shock – A study of students returning from youth expeditions. *Leisure Studies, 31*(4), 487–503.

Altbach, P. G., Reisberg, L., & Rumbley, L. E. (2009). *Trends in global higher education: Tracking an academic revolution.* A Report Prepared for the UNESCO 2009 World Conference on Higher Education.

Arnett, J. J. (2000). Emerging adulthood: A theory of development from the late teens through the twenties. *American Psychologist, 55*(5), 469–480.

Arnett, J. J. (2002). The psychology of globalization. *American Psychologist, 57*(10), 774–783.

Australian Government. (2014). *Research snapshot: Outgoing international mobility of Australian University students.* Canberra: Department of Education and Training.

Australian Government. (2016). *National strategy for international education 2025.* Canberra: Department of Education and Training.

Bagnoli, A. (2009). On 'An introspective journey': Identities and travel in young people's lives. *European Societies, 11*(3), 325–345.

Bandyopadhyay, S., & Bandyopadhyay, K. (2015). Factors influencing student participation in college study abroad programs. *Journal of International Education Research, 11*(2), 87–94.

Banks, J. A. (2004). Teaching for social justice, diversity, and citizenship in a global world. *The Educational Forum, 68*(4), 296–305.

Banks, J. A. (2009). Diversity and citizenship education in multicultural nations. *Multicultural Education Review, 1*(1), 1–28.

Baumeister, R. F. (1986). *Identity. Cultural change and the struggle for self.* New York, NY: Oxford University Press.

Beames, S. (2004). Overseas youth expeditions with Raleigh international: A rite of passage? *Australian Journal of Outdoor Education, 8*(1), 29–36.

Beck, U. (1992). *Risk society: Towards a new modernity.* London: Sage.

Buchan, N. R., Brewer, M. B., Grimalda, G., Wilson, R. K., Fatas, E., & Foddy, M. (2011). Global social identity and global cooperation. *Psychological Science, 22*(6), 821–828.

Butcher, J. (2011). Volunteer tourism may not be as good as it seems. *Tourism Recreation Research, 36*(1), 75–76.

Cairns, D. (2008). Moving in transition: Northern Ireland youth and geographical mobility. *Young, 16*(3), 227–249.

Cameron, J. D. (2013). Grounding experiential learning in 'thick' conceptions of global citizenship. In R. Tiessen & R. Huish (Eds.), *Globetrotting or global citizenship: Perils and potential of international experiential learning* (pp. 21–42). Toronto, ON: University of Toronto Press.

Casmir, F. L. (1984). Perception, cognition and intercultural communication. *Communication, 13*(1), 1–16.

Clarke, I., Flaherty, T. B., Wright, N. D., & McMillen, R. M. (2009). Student intercultural proficiency from study abroad programs. *Journal of Marketing Education, 31*(2), 173–181.

Côté, J., & Bynner, J. M. (2008). Changes in the transition to adulthood in the UK and Canada: The role of structure and agency in emerging adulthood. *Journal of Youth Studies, 11*(3), 251–268.

Crossman, J. E., & Clarke, M. (2010). International experience and graduate employability: Stakeholder perceptions on the connection. *Higher Education, 59*(5), 599–613.

Curran, S. J. (2007). The career value of education abroad. *International Educator, 16*(6), 48–49.

DallAlba, G., & Sidhu, R. (2015). Australian undergraduate students on the move: Experiencing outbound mobility. *Studies in Higher Education, 40*(4), 721–744.

Daly, A. (2011). Determinants of participating in Australian university student exchange programs. *Journal of Research in International Education, 10*(1), 58–70.

Davies, I., Evans, M., & Reid, A. (2005). Globalising citizenship education? A critique of 'global education' and 'citizenship education'. *British Journal of Educational Studies, 53*(1), 66–89.

De Cremer, D., & van Vugt, M. (1999). Social identification effects in social dilemmas. *European Journal of Social Psychology*, 29(7), 871–893.

Desforges, L. (2000). Traveling the world: Identity and travel biography. *Annals of Tourism Research*, 27(4), 926–945.

Diprose, K. (2012). Critical distance: Doing development education through international volunteering. *Area*, 44(2), 186–192.

Dolby, N. (2007). Reflections on nation: American undergraduates and education abroad. *Journal of Studies in International Education*, 11(2), 141–156.

Dower, N. (2000). The idea of global citizenship – A sympathetic assessment. *Global Society*, 14(4), 553–567.

Du Bois-Reymond, M. (1998). 'I don't want to commit myself yet': Young people's life concepts. *Journal of Youth Studies*, 1(1), 63–79.

Dwyer, M. M. (2004). More is better: The impact of study abroad program duration. *Frontiers: The Interdisciplinary Journal of Study Abroad*, 10, 151–163.

Elsrud, T. (2001). Risk creation in traveling: Backpacker adventure narration. *Annals of Tourism Research*, 28(3), 597–617.

Erikson, E. (1968). *Identity, youth and crisis*. New York, NY: Norton.

Erikson, E. H. (1980). *Identity and the life cycle*. London: W.W. Norton & Company.

Forsey, M., Broomhall, S., & Davis, J. (2012). Broadening the mind? Australian student reflections on the experience of overseas study. *Journal of Studies in International Education*, 16(2), 128–139.

Frändberg, L. (2014). Temporary transnational youth migration and its mobility links. *Mobilities*, 9(1), 146–164.

Franklin, K. (2010). Long-term career impact and professional applicability of the study abroad experience. *Frontiers: The Interdisciplinary Journal of Study Abroad*, 19, 169–190.

Frick, W. B. (1983). The symbolic growth experience. *Journal of Humanistic Psychology*, 23(1), 108–125.

Frick, W. B. (1987). The symbolic growth experience paradigm for a humanistic-existential learning theory. *Journal of Humanistic Psychology*, 27(4), 406–423.

Galston, W. A. (2001). Political knowledge, political engagement, and civic education. *Annual Review of Political Science*, 4(1), 217–234.

Geertz, C. (1973). *The interpretation of cultures*. New York, NY: Basic Books.

van Gennep, A. (1960). *The rites of passage*. Chicago, IL: The University of Chicago Press.

Gerhards, J., & Hans, S. (2013). Transnational human capital, education, and social inequality. Analyses of International Student Exchange. *Zeitschrift für Soziologie*, 42(2), 99–117.

Giddens, A. (1991). *Modernity and self-identity: Self and society in the late modern Age*. Stanford, CA: Stanford University Press.

Guttentag, D. A. (2009). The possible negative impacts of volunteer tourism. *International Journal of Tourism Research*, 11(6), 537–551.

Habermas, J. (2001). Why Europe needs a constitution. *New Left Review*, 11, 5–26.

Hadis, B. F. (2005). Why are they better students when they come back? Determinants of academic focusing gains in the study abroad experience. *Frontiers: The Interdisciplinary Journal of Study Abroad*, 11, 57–70.

Hall, E. T. (1959). *The silent language*. New York, NY: Doubleday & Company.

Hall, E. T. (1976). *Beyond culture*. New York, NY: Anchor Books/Doubleday.

Hall, E. T. (1983). *The dance of life: The other dimension of time*. New York, NY: Anchor Books, Doubleday.

Hibbert, J. F., Dickinson, J. E., & Curtin, S. (2013). Understanding the influence of interpersonal relationships on identity and tourism travel. *Anatolia: An International Journal of Tourism and Hospitality Research*, 24(1), 30–39.

Institute of International Education. (2007). *Open doors report on international educational exchange*. 2007 Fast Facts: International Students in the U.S.

Institute of International Education. (2015). *Open doors report on international educational exchange*. 2015 Fast Facts: International Students in the U.S.

Jorgenson, S. (2010). De-centering and re-visioning global citizenship education abroad programs. *International Journal of Development Education and Global Learning*, 3(1), 23–38.

Kasravi, J. (2009). *Factors influencing the decision to study abroad for students of color: Moving beyond the barriers* (Doctoral thesis). University of Minnesota.

King, A. (2011). Minding the gap? Young people's accounts of taking a Gap year as a form of identity work in higher education. *Journal of Youth Studies*, 14(3), 341–357.

Kofman, E. (2005). Citizenship, migration and the reassertion of national identity. *Citizenship Studies*, 9(5), 453–467.

Landon, A. C., Tarrant, M. A., Rubin, D. L., & Stoner, L. (2017). Beyond 'just do it': Fostering higher-order learning outcomes in short-term study abroad. *AERA Open*, 3(1), 1–7.

Larsen, S. (2007). Aspects of a psychology of the tourist experience. *Scandinavian Journal of Hospitality and Tourism*, 7(1), 7–18.

Larson, R. W. (2000). Toward a psychology of positive youth development. *American Psychologist*, 55(1), 170–183.

Lewin, R. (2009). Transforming the study abroad experience into a collective priority. *Peer Review*, 11(4), 8–11.

Liang, K., Caton, K., & Hill, D. J. (2015). Lessons from the road: Travel, lifewide learning, and higher education. *Journal of Teaching in Travel & Tourism*, 15(3), 225–241.

Lincoln Commission. (2005). *Global competence and national needs: One million Americans studying abroad*. Washington, DC: Commission on the Abraham Lincoln Fellowship Program.

Lyons, K., Hanley, J., Wearing, S., & Neil, J. (2012). Gap year volunteer tourism: Myths of global citizenship? *Annals of Tourism Research*, 39(1), 361–378.

Lyons, K. D. (2015). Reciprocity in volunteer tourism and travelism. In T. V. Singh (Ed.), *Challenges in tourism research* (pp. 106–112). Bristol: Channel View Publications.

Markus, H. R., & Kitayama, S. (1991). Culture and the self: Implications for cognition, emotion, and motivation. *Psychological Review*, 98(2), 224–253.

Marsella, A. J. (1985). Culture, self, and mental disorder. In A. J. Marsella, G. DeVos, & F. L. K. Hsu (Eds.), *Culture and self: Asian and western perspectives* (pp. 281–307). New York, NY: Tavistock Publications.

M'Balia, T. (2013). The problematization of racial/ethnic minority student participation in U.S. study abroad. *Applied Linguistics Review*, 4(2), 365–390.

McBride, A. M., Brav, J., Menon, N., & Sherraden, M. (2006). Limitations of civic service: Critical perspectives. *Community Development Journal*, 41(3), 307–320.

Mehmetoglu, M., & Engen, M. (2011). Pine and Gilmore's concept of experience economy and its dimensions: An empirical examination in tourism. *Journal of Quality Assurance in Hospitality & Tourism, 12*(4), 237–255.

Minh-ha, T. T. (1994). Other than myself/my other self. In G. Robertson, M. Mash, L. Tickner, J. Bird, B. Curtis, & T. Putnam (Eds.), *Travellers' tales: Narratives of home and displacement* (pp. 9–26). London: Routledge.

Modood, T. (2007). *Multiculturalism. A civic idea.* Cambridge: Polity.

Montuori, A., & Fahim, U. (2004). Cross-cultural encounter as an opportunity for personal growth. *Journal of Humanistic Psychology, 44*(2), 243–265.

Moon, J. A. (2004). *A handbook of reflective and experiential learning: Theory and practice.* New York, NY: Routledge.

Morais, D. B., & Ogden, A. C. (2011). Initial development and validation of the global citizenship scale. *Journal of Studies in International Education, 15*(5), 445–466.

Mowforth, M. & Munt, I. (2015). *Tourism and sustainability: development, globalisation and new tourism in the third world.* London: Routledge.

Murray Brux, J., & Fry, B. (2010). Multicultural students in study abroad: Their interests, their issues, and their constraints. *Journal of Studies in International Education, 14*(5), 508–527.

Noddings, N. (2005). Global citizenship: Promises and problems. In N. Noddings (Ed.), *Educating citizens for global awareness* (pp. 1–21). New York, NY: Teachers College Press.

Nussbaum, M. C. (1996). *For love of country?* (J. Cohen, Ed.), Boston, MA: Beacon Press.

Olsen, A. (2008). International mobility of Australian university students: 2005. *Journal of Studies in International Education, 12*(4), 364–374.

Ozdowski, S. (2012). *Australian multiculturalism: The roots of its success.* Third international conference on human rights education: Promoting change in times of transition and crisis. Krakow: The Jagiellonian University in Krakow.

Pine, B. J., & Gilmore, J. H. (1999). *The experience economy.* Boston, MA: Harvard Business School Press.

Reau, B. (2012, July). Grand tour with a degree (C. Goulden, Trans.). *Le Monde Diplomatique,* p. 14.

Reysen, S., & Katzarska-Miller, I. (2013). A model of global citizenship: Antecedents and outcomes. *International Journal of Psychology, 48*(5), 858–870.

Riley, P. J. (1988). Road culture of international long-term budget travelers. *Annals of Tourism Research, 15*(3), 313–328.

Rubin, D. L., Landon, A., Tarrant, M., Stoner, L., & Mintz, L. (2016). Measuring attitudes toward the rights of indigenous peoples: An index of global citizenship. *Journal of Global Citizenship & Equity Education, 5*(1), 101–116.

Salisbury, M. H., Paulsen, M. B., & Pascarella, E. T. (2010). To see the world or stay at home: Applying an integrated student choice model to explore the gender gap in the intent to study abroad. *Research in Higher Education, 51*(7), 615–640.

Schattle, H. (2009). Global citizenship in theory and practice. In R. Lewin (Ed.), *The handbook of practice and research in study abroad: Higher education and the quest for global citizenship* (pp. 3–18). New York, NY: Routledge.

Shulman, S., Blatt, S. J., & Walsh, S. (2006). The extended journey and transition to adulthood: The case of Israeli backpackers. *Journal of Youth Studies, 9*(2), 231–246.

Shulman, S., Feldman, B., Blatt, S. J., Cohen, O., & Mahler, A. (2005). Emerging adulthood: Age-related tasks and underlying self processes. *Journal of Adolescent Research, 20*(5), 577–603.

Simpson, K. (2004). 'Doing development': The gap year, volunteer-tourists and a popular practice of development. *Journal of International Development, 16*(5), 681–692.

Smith, D., Smith, M., Robbins, K., Eash, N., & Walker, F. (2013). Traditionally under-represented students' perceptions of a study abroad experience. *NACTA Journal, 57*(3a), 15–20.

Starr-Glass, D. (2016). Repositioning study abroad as a rite of passage: Impact, implications and implementation. In D. M. Velliaris & D. Coleman-George (Eds.), *Study abroad programs and outward mobility* (pp. 89–114). Hershey, PA: IGI Global.

Stoner, K. R., Tarrant, M. A., Perry, L., Stoner, L., Wearing, S., & Lyons, K. (2014). Global citizenship as a learning outcome of educational travel. *Journal of Teaching in Travel & Tourism, 14*(2), 149–163.

Sussman, N. M. (2000). The dynamic nature of cultural identity throughout cultural transitions: Why home is not so sweet. *Personality & Social Psychology Review, 4*(4), 355–373.

Sutton, R. C., & Rubin, D. L. (2004). The GLOSSARI project: Initial findings from a system-wide research initiative on study abroad learning outcomes. *Frontiers: The Interdisciplinary Journal of Study Abroad, 10,* 65–82.

Sutton, R. C., & Rubin, D. L. (2010). *Documenting the academic impact of study abroad: Final report of the GLOSSARI project.* Annual conference of NAFSA: Association of International Educators, Kansas City, MO.

Tajfel, H. & Turner, J. (1979). An integrative theory of intergroup conflict. In W. G. Austin & S. Worchel (Eds.), *The social psychology of intergroup relations* (pp. 33–47). Monterey, CA: Brooks/Cole.

Tarrant, M. A. (2010). A conceptual framework for exploring the role of studies abroad in nurturing global citizenship. *Journal of Studies in International Education, 14*(5), 433–451.

Tarrant, M. A., Lyons, K., Stoner, L., Kyle, G. T., Wearing, S., & Poudyal, N. (2014). Global citizenry, educational travel and sustainable tourism: Evidence from Australia and New Zealand. *Journal of Sustainable Tourism, 22*(3), 403–420.

Thomson, R., & Taylor, R. (2005). Between cosmopolitanism and the locals: Mobility as a resource in the transition to adulthood. *Young, 13*(4), 327–342.

Tinto, V. (1988). Stages of student departure: Reflections on the longitudinal character of student leaving. *The Journal of Higher Education, 59*(4), 438–455.

Towner, J. (1985). The grand tour. *Annals of Tourism Research, 12*(3), 297–333.

Trooboff, S., Vande Berg, M., & Rayman, J. (2008). Employer attitudes toward study abroad. *Frontiers: The Interdisciplinary Journal of Study Abroad, 15,* 17–33.

Twombly, S. B., Salisbury, M. H., Tumanut, S. D., & Klute, P. (2012). *Study abroad in a new global century: Renewing the promise, refining the purpose, ASHE higher education report.* Hoboken, NJ: John Wiley & Sons.

UNWTO. (2016). UNWTO tourism highlights. 2016 Edition.

Ward, C. (2004). Psychological theories of culture contact and their implications for intercultural training and interventions. In D. Landis, J. M. Bennett, & M. J. Bennett (Eds.), *Handbook of intercultural training* (3rd ed., pp. 185–216). Thousand Oaks, CA: Sage.

Wearing, S. (2002). Re-centering the self in volunteer tourism. In G. M. S. Dann (Ed.), *The tourist as a metaphor of the social world* (pp. 237–262). New York, NY: CABI.

Wearing, S., Tarrant, M. A., Schweinsberg, S., Lyons, K., & Stoner, K. (2015). Exploring the global in student assessment and feedback for sustainable tourism education. In G. Moscardo & P. Benckendorff (Eds.), *Education for sustainability in tourism: A handbook of processes, resources, and strategies* (pp. 101–115). New York, NY: Springer.

Weaver, D., & Lawton, L. (2010). *Tourism management* (4th ed). Milton: John Wiley & Sons Australia.

Wick, D. J. (2011). *Study abroad for students of color: A third space for negotiating agency and identity* (Doctoral thesis). San Francisco State University.

Williams, T. R. (2005). Exploring the impact of study abroad on students' intercultural communication skills: Adaptability and sensitivity. *Journal of Studies in International Education, 9*(4), 356–371.

Wilson, J., Fisher, D., & Moore, K. (2009). The OE goes 'home': Cultural aspects of a working holiday experience. *Tourist Studies, 9*(1), 3–21.

Yoon, K. (2014). Transnational youth mobility in the neoliberal economy of experience. *Journal of Youth Studies, 17*(8), 1014–1028.

Cosmopolitan empathy in volunteer tourism: a psychosocial perspective

Émilie Crossley

ABSTRACT
Volunteer tourism provides a means of proximate engagement with usually distant others, emphasising reciprocity, cultural learning and humanitarianism in poor communities. As such, the practice has come to be investigated for its potential to engender global citizenship, a broader scope of emotional identification, and new kinds of progressive transnational social spaces. This paper focuses on the intersection between volunteer tourism and cosmopolitan empathy, outlining an account of cosmopolitan empathy that draws on a Lacanian psychosocial reading of tourist subjectivity. This theorisation conceptualises cosmopolitan empathy as an emergent property of interrelated social and psychic fields, which results in the affect serving both ideological and psychological functions. I argue that bridging geographical distance through travel presents volunteer tourists with encounters that can potentially destabilise the discourses and fantasies of the needy, grateful Other underpinning their experiences of cosmopolitan empathy, thus disrupting the conventional spatial ontology of affect that frequently dominates theoretical discussions of cosmopolitanism.

Introduction

Volunteer tourism aims to provide small-scale, ecologically sensitive experiences that allow tourists to interact with visited communities in a reciprocal, mutually beneficial way (Wearing, 2001). Advocates of volunteer tourism suggest that it enables participants to live alongside beneficiaries of community development or humanitarian projects, thereby providing an unusual means of proximate engagement with usually distant global others. Unlike most forms of ethical consumption, volunteer tourists as end-consumers are brought into direct contact with those supposedly benefiting from their ethical travel choices (Sin, 2010). Volunteer tourism is framed as a practice that transforms imaginings about the suffering of distant others into proximate acts of help and care. As such, volunteer tourism has come to be investigated for its potential to foster cosmopolitan empathy (Beck, 2006; Swain, 2009), new geographies of responsibility, care and compassion (Massey, 2004; Mitchell, 2007; Mostafanezhad, 2013a; Popke, 2006, 2007; Sin, 2010) and the development of ethical subjectivity or global citizenship (Baillie Smith & Laurie, 2011; Butcher & Smith, 2015; Crossley, 2014; Diprose, 2012; Heron, 2011; Lyons, Hanley, Wearing, & Neil, 2012; McGehee & Santos, 2005; Tiessen & Epprecht, 2012).

In this paper, I develop a theoretical understanding of cosmopolitan empathy as it relates to volunteer tourism. What is of particular interest to me is whether cosmopolitan empathy can be conceived of as a motivation for people to volunteer abroad, the sincerity of this affective response to distant others, and whether such empathy endures beyond the volunteer tourism experience. I also address how cosmopolitan empathy is oriented towards the suffering of particular social groups deemed worthy of volunteer tourists' concern. In order to achieve this, I explore the potential of Lacanian-inflected psychosocial studies to enhance our understanding of cosmopolitan empathy in volunteer tourism, thereby contributing to an emerging critical literature on emotion and affect in tourism (Crossley, 2012a, 2012b, 2014; Molz, 2015; Picard & Robinson, 2012; Tucker, 2009) and empathy more specifically (Mostafanezhad, 2014; Tucker, 2016). Psychosocial studies theorises subjectivity non-dualistically as an emergent property of interconnected social and psychic fields, avoiding the reductive, essentialising tendencies that make psychology off-putting for some critical tourism researchers (McCabe, 2005; Moore, 2002). This approach resists the pervasive characterisation of tourists as 'rational, wholly conscious, and psychically-integrated individuals' (Kingsbury, 2005, p. 114) and places an emphasis on affect, emotion and unconscious dynamics. In the sections that follow, I draw on Lacanian psychosocial theory to explore processes of identification and unconscious affects in order to enrich our theoretical understanding of cosmopolitan empathy in volunteer tourism.

Volunteer tourism and discourses of global citizenship

Wearing (2001, p. 240) defines volunteer tourism as a practice in which people 'volunteer in an organized way to undertake holidays that may involve aiding or alleviating the material poverty of some groups in society, the restoration of certain environments, or research into aspects of society or environment'. It is those forms of volunteer tourism purporting to alleviate material poverty that provide volunteer tourists with the most direct link to distant others and which, therefore, are most relevant to an exploration of cosmopolitan empathy conceived of as a response to the suffering of distant others. Before examining theorisations of cosmopolitan empathy and the form that it takes in volunteer tourism, I want to take a broader view of the practice in order to acknowledge the multiple ways in which volunteer tourism is structured and framed by discourses of global citizenship, ethical consumerism and transformative learning. I argue that these discourses – emanating from sources such as the media, tourism marketing, education and government policy – present volunteer tourists with a series of imperatives regarding how they are supposed to act, feel and change throughout their travel experience. These discourses construct an expectation for volunteer tourists to undergo significant personal growth and transformation, become more cosmopolitan and charitable, and experience challenging yet beneficial emotions. This context is therefore important to examine in relation to cosmopolitan empathy in volunteer tourism and has the potential to elucidate how empathy arises and the precise form that it takes through these tourism practices.

Volunteer tourism is framed by discourses that promote an educational model of travel and the development of a more 'global' worldview that is often referred to as 'global citizenship' (Baillie Smith & Laurie, 2011; Butcher & Smith, 2015; Crossley, 2014; Diprose, 2012; Heron, 2011; Lyons et al., 2012; Tiessen & Epprecht, 2012). Butcher and Smith (2015, p. 99) argue that global citizenship represents an 'outsourcing of citizenship' from the national to the global level, whereby the 'Global South' becomes a platform for affluent Western subjects to explore ethical subjectivity through practices such as volunteer tourism. Butcher and Smith (2015, p. 97) also suggest that global citizenship is framed by discourses of consumption that often present 'global action in terms of what an individual can do in the context of their daily life, through consumption and lifestyle'. The advent of ethical consumerism has forced difficult questions to be asked about the evolving nature of subjectivity in Western societies, distinctions between citizens and consumers, and the avenues available to subjects for ethical and political action. For example, contrary to arguments that pit individualistic 'consumers' again collectively oriented 'citizens', Clarke, Barnett, Cloke, and Malpass (2007) suggest that 'new forms of *citizenly action* are currently being configured through creative redeployment of the repertoires of consumerism' (p. 5, original emphasis; see also Barnett, Cloke, Clarke, & Malpass, 2005, 2010). Furthermore, the blurring of public and private spaces brought about by modern-day consumerism is implied to contain the seeds of consumerism's redemption through the 'transnationalisation of responsibilities' and fostering of more cosmopolitan sensibilities amongst consumers (Clarke et al., 2007, p. 23; Mitcheletti, 2003). This is particularly the case for volunteer tourism, in which links between consumption and distant people and places become tangible and enacted through the act of travel, and where the development of global citizenship is promoted as one of its aims.

I argue that this framing of 'global action' in terms of lifestyle and consumerism has had an individualising effect on how agency and change are conceptualised in relation to global socio-economic development. In the context of volunteer tourism, this has located the individual volunteer tourist as the agent of change and has led to a proliferation of discourses of personal growth, transformative learning and moral self-development, for which we also find substantial empirical evidence from volunteer tourists' narratives (Coghlan & Gooch, 2011; Crossley, 2012b; McGehee & Santos, 2005; Zahra, 2011; Zahra & McIntosh, 2007). Wearing and Neil (2000) suggest that it is in part volunteer tourists' interactions with others from different cultural backgrounds that can bring about a renegotiation of identity and impact on the tourist's self so profoundly. Coghlan and Gooch (2011) make a similar argument for politicised learning in volunteer tourism, suggesting that the practice can be interpreted as a form of 'transformative learning' in which the tourist's social position and naturalised ideologies are critically re-evaluated through shared experiences with others (both other volunteer tourists and local people in the tourism destination). While the claim that identity and sense of self can alter through travel is certainly not specific to volunteer tourism, what sets the practice apart is this vision of ethical and politicised change that it presents (Butcher & Smith, 2015; Lyons et al., 2012).

Discourses of self-change, transformative learning and global citizenship are strong across a range of international volunteering practices, including volunteer tourism and the related activity of international service learning, which is run through programmes in many Western universities, particularly in countries such as the USA and Australia (Annette, 2002; Kiely, 2004; Lyons & Wearing, 2008; Tiessen & Epprecht, 2012). The growth of international service learning, which entails many of the same activities as volunteer tourism, albeit undertaken within a more structured, reflective and pedagogical framework, can be seen as reflecting a globalisation of higher education in which universities are under pressure to produce graduates with a 'global perspective' (Appiah, 2008; Heron, 2011; Osler & Starkey, 2003; Porter & Monard, 2001; Standish, 2012). Crabtree (2008, p. 18) describes the mission of international service learning as 'increasing participants' global awareness and development of humane values, building intercultural understanding and communication, and enhancing civic mindedness and leadership skills.' This interrelation between 'global awareness' and 'civic mindedness' is well captured by the concept of global citizenship. In addition, international service learning claims to promote concern for suffering and injustice occurring in geographically distant places (Heron, 2011; Kiely, 2004) and claims to move students from 'a charity orientation toward more of a social justice orientation' (Crabtree, 2008, p. 26).

What is interesting in comparing international service learning with volunteer tourism is the relative emphasis placed in the former on the development of a political consciousness and solidarity with global others. In contrast, volunteer tourism's more charitable ethos has been criticised for producing depoliticised responses to poverty that 'preclude political reflection and prevent a collective model for social justice that can transcend cultural particulars' (Vrasti, 2013, p. 57). Rather than empowering and enabling tourists to contribute meaningfully to development projects, Butcher and Smith (2015, p. 100) argue that volunteer tourism narcissistically 'leads away from addressing the pressing material needs of others in the context of their lives and towards addressing the crisis of political identity in the West'. In this context, the emergence of cosmopolitan empathy could cynically be read as a willed, self-indulgent emotional experience that is of greater benefit to the volunteer tourist than to any member of the visited community (Crossley, 2012a, 2012b). Having said this, there is empirical evidence to suggest that volunteer tourism can serve as a mechanism for consciousness-raising, can increase support for social movements and encourage participation in political activism (McGehee, 2002; McGehee & Norman, 2001; McGehee & Santos, 2005).

Further critiques of volunteer tourism suggest that rather than engendering cross-cultural understanding and global interconnectedness, the practice can actually reinforce cultural stereotypes and 'Othering' (Simpson, 2004; Snee, 2013; Wearing & Wearing, 2006) and encourage Eurocentric attitudes (Palacios, 2010). Raymond and Hall (2008) report that volunteer tourists can perceive positive interactions with visited others as being an 'exception to the rule' and treat these encounters as providing fond memories rather than instigating lasting friendships. Simpson (2005, 2004), in her work on the British pre-university gap year, points to a shallowness of understanding that is brought away by volunteer tourists, including tendencies to romanticise poverty. Mostafanezhad (2013b) similarly levels criticism at the non-reciprocal relation of care advanced by volunteer tourism, which she sees as embodied in a 'humanitarian gaze' that provides a discourse for imagining helping relationships and scripting the roles of who is for saving and who is the saviour in the tourism encounter. Mostafanezhad argues that this missionary-style gaze sets up a binary and hierarchy between care givers and receivers that maintains a paternalistic power dynamic. As Sin (2010, p. 985) points out, the problem inherent in forming relationships of care is that they 'immediately imply the lack of equal relationships since the carer naturally assumes the position of privilege and power'. The emphasis placed on responsibility and care in volunteer tourism has consequently been critiqued as normalising a dynamic of activity/passivity in which the volunteer is assumed to be the natural active, moral agent (Barnett & Land, 2007; Silk, 2004; Sin, 2010).

This scripting of volunteer tourism roles within relationships of care connects to research that has elucidated how a particular image of the 'Third World Other' has come to be discursively framed as the proper locus of Western subjects' concern through media imagery, travel brochures, and volunteer tourists' narratives (Mostafanezhad, 2013a, 2014; Sin, 2010; Vrasti, 2013). More specifically, Mostafanezhad (2013a) argues that vulnerable children in developing countries constitute the predominant object of compassion and empathy in volunteer tourism, revealing another dimension to an already uneven power dynamic between tourists and members of visited communities. Further studies have revealed the centrality of volunteer tourists' own socioeconomic status in feeling capable or inclined to undertake voluntary work, constructing a narrative that associates 'doing well in life' with wanting to 'give back' to disadvantaged communities (Brown & Lehto, 2005, p. 488; Zahra & McIntosh, 2007). The notion of 'giving something back', which is connected to a similar discourse of needing to 'make a difference' through ethical travel (Butcher & Smith, 2010; Raymond, 2008; Simpson, 2004; Vrasti, 2013), demonstrates the highly contextual nature of altruistic expressions, locating volunteer tourism as a potentially reparative act on the part of the global rich towards the global poor.

Sinervo (2011) refers to this entanglement of discourses, affectivity and ethical frameworks as the 'moral economy' of volunteer tourism, which produces particular expectations and obligations for both tourists and local people in visited communities. In this section, I have begun to explore how this moral economy functions by reviewing the multifarious ways in which volunteer tourism is structured and framed by discourses of global citizenship, ethical consumerism, transformative learning and personal growth. The empirical evidence available suggests that volunteer tourism is fraught with contradictions and balances precariously between doing good and doing harm in relation to the issues of global poverty and cross-cultural communication. I now want to extend this analysis by focusing on the concept of cosmopolitan empathy in order to provide a critical perspective on questions that are central to volunteer tourism, including the part played by affects and emotions in motivating people to participate in volunteer tourism and the potential for the practice to engender a lasting disposition of care and empathy towards global others.

Cosmopolitanism, empathy and care

As I have shown, volunteer tourism is a practice that is deeply influenced by discourses of global citizenship and can thus be thought of as a 'global phenomenon deeply enmeshed in the flows, forces and imaginings of globalization' (Swain, 2009, p. 508). I now want to delve deeper into the theory of global citizenship and cosmopolitanism as a way of leading into a discussion of cosmopolitan empathy in volunteer tourism. Volunteer tourism is rooted in the reconfiguration of space brought about by globalisation, through which spatially distant groups of people now regularly interface, be it physically or virtually. Giddens notes that, '[i]n high modernity, the influence of distant happenings on proximate events, and on intimacies of the self, become more and more commonplace' (1991, p. 4). Cosmopolitanism has been used as a theoretical lens through which this new spatial and affective configuration can be understood. Cosmopolitanism, or global citizenship, can be conceptualised as an identity or sense of belonging to an imagined global community that supersedes national citizenship (Appiah, 2010; Beck, 1998; Carter, 2004; Diprose, 2012; Skrbis, Kendall, & Woodward, 2004; Vertovec & Cohen, 2002). As such, cosmopolitanism as a normative concept proposes that citizens freed from the divisive confines of national citizenship will be unified by their common humanity, leading to a global society in which cultural diversity is embraced and geographically distant others are treated as neighbours (Carter, 2004; Lyons et al., 2012). A cosmopolitan disposition is commonly associated with qualities such as openness to cultural difference, semiotic proficiency in reading other cultures, lack of geographical rootedness, and tastes or attitudes that reflect globalised consumption patterns (Appiah, 2010; Beck, 2006; Hannerz, 2004; Skrbis et al., 2004; Swain, 2009; Vertovec & Cohen, 2002). Cosmopolitanism can therefore refer equally to identity, mobility, citizenship, consumption, responsibility and ethics.

Two competing theoretical perspectives have emerged within the literature on cosmopolitanism. Lisle (2009) refers to the first position as 'progressive' cosmopolitanism, associated with the 'cosmopolitan manifesto' advocated by Beck (1998), which conceptualises subjects as self-reflective individuals and promotes the development of universalistic ethical norms and codes of practice towards the end of global emancipation. 'Critical' cosmopolitanism, on the other hand, has emerged as a feminist critique of the dominant discourse of cosmopolitanism, criticising it for presenting a masculinist version of individualised subjectivity and abstract, disembedded systems of ethics and rights (cf. Anderson, 2001; Brassett & Bulley, 2007; Calhoun, 2002; Cheah & Robbins, 1998; Delanty, 2006; Derrida, 2001; Erskine, 2002; Linklater, 2007; Lisle, 2009; McRobbie, 2006; Mendieta, 2009; Mitchell, 2007; Molz, 2005; Robinson, 1999; Swain, 2009; Vertovec & Cohen, 2002). What arises from this second vision of cosmopolitanism is an awareness of the importance of relationality, intersubjectivity, and what Judith Butler sees as an endless undertaking of subjects being '(re)constituted through dialectical processes of recognition, within multiple networks of power' (Mitchell, 2007, p. 711). So, while conventional cosmopolitanism advocates an ethics based on universal rights and responsibilities, its critical variant is founded upon a feminist ethics of care that conceives of ethics as something that cannot be set out as abstract principles but which emerges through specific sites and social relationships producing the need for care (Gilligan, 1982; Held, 1993, 1995; Jagger, 1989, 1991, 1995; Koehn, 1998; Lawson, 2007; Tronto, 1993). Nevertheless, this critical perspective still relies on a theoretical foundation that advocates care as a guiding ethical principal.

Care and empathy are often under-theorised or glossed over in theoretical discussions of cosmopolitanism, which tend to eschew analysis of emotion or affect in favour of cosmopolitanism's more overtly political facets such as identity, mobility, citizenship or consumption (Beck, 1998; Vertovec & Cohen, 2002). However, Skrbis et al. (2004, p. 127) present empathy as an essential component of cosmopolitanism, stressing that a cosmopolitan disposition 'must involve emotional and moral/ethical commitments' and that such commitments are 'demonstrated by an empathy for and interest in other cultures'. Beck (2006, p. 6) suggests that there has been a 'globalisation of emotions' in which 'the spaces of our emotional imagination have expanded in a transnational sense'. This has enabled what he terms 'cosmopolitan empathy' to arise, which refers to people's ability to adopt the perspective of geographically distant others and to feel compassion for their plight; to put ourselves in the place of victims of war, famine, drought and so on. As Kyriakidou (2008, p. 159) puts it, cosmopolitan empathy 'describes the infiltration of people's everyday local experiences and moral lifeworlds with emotionally engaging values that orient them towards the global and geographically distant others'. Chouliaraki (2006) describes this process as 'the closing of moral distance' that has resulted from new physical and virtual connections between people across space.

Cosmopolitan empathy suggests that this new social interconnectedness and cultural hybridity brought about by processes of globalisation has impacted on not only how citizens conceptualise their place in global society, but also on how they feel about their connection to others. As a result, geographers have begun to question commonly held assumptions about spatial ontologies of feeling and affect. Callon and Law (2004, p. 3) describe conventional, naturalised understandings of space as a 'fiction of "natural space"', while Mitchell (2007, p. 706) complains of the dominance of formulaic linkages between '[d]istance = coldness and nearness = warmth' in theoretical discussions about cosmopolitanism, which would seem to preclude the possibility of extensions of care or compassion towards distant others. This resonates with Thien's (2005, p. 193) discussion of intimacy, which, she says, 'assumes a distance covered, a space traversed to achieve a desired familiarity with another. As a vision/version of an achieved relationship (self to other), it is the antithesis of distance'. Similarly, Massey's (2004, p. 9) work disrupts the conventional opposition between space and place – in which space is theorised as abstract whereas place is seen as the site of social relationships and meaning – that has reinforced what she refers to as a 'locally centred, Russian doll geography of care and responsibility', thereby opening up global space to novel analyses of affect.

Höijer (2004, p. 514) situates cosmopolitan empathy as part of a broader sensibility of global compassion produced by political discourse, humanitarian organisations and the media, pointing, in particular, to the extensive media coverage of global conflicts and crises that bring images of distant suffering into the homes of citizens living in more affluent or stable parts of the world. Silk (1998) refers to such exposure to images through the media as 'mediated quasi-interactions', which he lists alongside 'face-to-face interactions' and 'mediated interactions' (such as chatting to someone online) as structures of interaction that can facilitate caring at a distance. He also draws on Smith's (1998) distinction between benevolence, caring *about* others, and beneficence, caring *for* others, stressing that the former need not always precede the latter, as in the case of enacting care out of a sense of guilt (Silk, 2000, see also 2004; Smith, 2000). These theoretical threads are useful for situating volunteer tourism, which initially involves mediated quasi-interactions with others through advertisements, charity appeals and the news, and later face-to-face interactions, during which care is enacted through voluntary work. Therefore, in volunteer tourism benevolence, which is akin to the concept of cosmopolitan empathy in this context, occurs at a distance whereas beneficence is enacted proximately.

Empirical studies of cosmopolitan empathy have tended to focus on responses to media coverage of humanitarian and natural disasters, and it is worth reflecting on some of their findings here. In a study of Greek citizens' responses to recent natural disasters, Kyriakidou (2008) observed that people found it difficult to experience an intense emotional response when considering a group of victims, in contrast to the power of individual human stories. She also found that cosmopolitan empathy appeared to be banal – 'an unreflective, taken-for-granted sentiment' that could quickly be moved on from (Kyriakidou, 2008, p. 163). Similarly, Höijer's (2004) interviews with Swedish and Norwegian citizens about news coverage of the Kosovo War found that viewers' compassion or empathy was dependent on the power of visual imagery and the presence of 'ideal victim images'. Her respondents were drawn to images of children, elderly people and women, who were perceived as more vulnerable than men caught up in the conflict and thus more deserving of their compassion. These empirical findings connect strongly to Mostafanezhad's (2013a) observation that vulnerable children in developing countries constitute the predominant object of compassion and empathy in volunteer tourism.

Mostafanezhad (2014, p. 70) describes cosmopolitan empathy as 'an emotional response to the plight of the poor in the Global South' that serves as 'a corollary of the cultural logic and economic policies of neoliberal global capitalism' (p. 86). However, beyond this reference to emotion, the concept of cosmopolitan empathy remains relatively untheorised in Mostafanezhad's work and, I argue, warrants a more in-depth theoretical exploration. More broadly, Tucker (2016) notes the lack of attention that has been paid to the concept of empathy, in any form, within the tourism literature despite its potential to articulate the affective dynamics at play within a variety of contexts. Indeed, in relation to volunteer tourism, the term 'cosmopolitan empathy' has not to date been used beyond Mostafanezhad's work (see also Swain, 2016), although 'empathy' does make an appearance in the literature, usually in reference to tourists' development of cross-cultural competencies once in the volunteer tourism site (cf. Lyons et al., 2012; Mostafanezhad, 2013b). Instead, researchers have tended to frame their enquiries into volunteer tourists' feelings towards distant others in terms of altruism, sympathy or compassion. It is often unclear in the broader literature on tourism and cosmopolitanism whether what is referred to as 'cosmopolitan empathy' is any different from these related affects. Crucially, empathy implies the capacity to imaginatively put oneself in the place of another and, in some sense, to feel what they feel. Sympathy, on the other hand, can be experienced without this perspective taking. I argue, therefore, that there is a need for greater conceptual clarity and theoretical cogency if cosmopolitan empathy is to be utilised in the analysis of volunteer tourism. In the sections that follow, I lay the groundwork for a Lacanian-inflected psychosocial approach to cosmopolitan empathy that I believe can go some way to provide the concept with greater explanatory power.

Lacanian psychosocial theory and tourism studies

The recent emergence of what has come to be known as 'psychosocial studies' (Frosh, 2003) reflects a growing insistence within the social sciences upon theorising subjectivity in ways that privilege neither its social nor psychological dimensions. Psychosocial studies can be conceptualised as a theoretically and methodologically plural field aiming to articulate 'a place of "suture" between elements whose contribution to the production of the human subject is normally theorised separately' (Frosh & Baraitser, 2008, p. 348). Examples of such pervasive theoretical dualisms include individual/society, psyche/social, structure/agency and body/mind. Psychosocial researchers claim that mainstream psychology disavows forms of intersubjectivity and relationality that bind subjects socially, producing a reductive and ideologically distorted theorisation of the subject (Blackman, Cromby, Hook, Papadopoulos, & Walkerdine, 2008; Henriques, Hollway, Urwin, Venn, & Walkerdine, 1984). Some note how psychology has been complicit in perpetuating neoliberal governance by portraying people as rational individuals, unconstrained by social ties, who ultimately look out for their own interests (Parker, 2007; Parker & Shotter, 1990). Critical psychology and psychosocial research therefore emanates not only from concerns regarding methodological accuracy, but also from a commitment to an emancipatory politics (Hayes, 2001; Hepburn, 2003; Hook, 2005; Parker, 2002). Frosh (2003) lists some of the theoretical elements contributing to psychosocial studies as psychoanalysis, systems theory, discourse analysis, feminist theory and phenomenology, to which we could add the theoretical work relating to affect (not exclusively psychoanalytic in nature) that has gained increasing prominence in recent years (Brennan, 2004; Clough & Halley, 2007; Gregg & Seigworth, 2010). In this paper, I focus on psychoanalytic psychosocial studies, drawing in particular on Lacanian theory.

I argue that psychoanalytic theory provides a theoretical lexicon that is unparalleled in its sophistication in dealing with unconscious, affective and conflicting elements of subjectivity. While much of this complexity finds discursive expression, Frosh (2001, p. 29) reminds us that 'there is a point at which discourse fails', in that there may be meaning that can only be retrospectively put into words and even then this expression may only be partial or faltering against the spectre of something that remains forever outside of symbolisation. It is this capability of psychoanalysis to deal with extra-discursive aspects of subjectivity and the complexity of personal, emotional worlds that has led to it finding favour among psychosocial researchers who see purely discursive accounts of subjectivity as lacking in explanatory power. Psychosocial studies place a strong empirical emphasis upon lived experience and personal narratives, and have been typified by the use of in-depth qualitative interviews to generate data (Walkerdine, 2008). In this context, psychoanalysis has been utilised as a strategy for 'enriching' and 'thickening' discursive interpretations of such narratives, allowing researchers to look for conscious and unconscious 'reasons' behind subjects' investments[1] in particular subject positions (Frosh & Saville Young, 2008). This has allowed psychoanalytic psychosocial researchers to propose that theirs is a 'why' rather than merely a 'how' approach to the discursive construction of subjectivity, providing insights into why subjects appropriate particular discursive resources out of an array of possibilities (Frosh, Phoenix, & Pattman, 2003).

It is claimed that by paying attention to biographically derived unconscious dynamics we can better understand this 'stickiness' of identities (Edley, 2006) and why one becomes positioned in ways that may induce internal conflicts or contradictions; ways that do not seem rational but may nonetheless make sense on some unconscious or emotional level.

Lacanian theory has appealed to psychosocial researchers wishing to forge a synthesis between discursive psychology and psychoanalysis. Through its constitution in language and Otherness, the Lacanian subject is effectively decentred, challenging the existence of 'stable, causal mental structures that can be known through interpretative practices' and instead introducing instability, uncertainty and scepticism regarding any 'true meaning' lying at the heart of psychic reality (Frosh & Emerson, 2005, p. 308). This leads Frosh and Emerson to classify Lacanian theory as an example of 'postmodern' psychoanalysis, in contrast to the more 'humanistic' strands associated with Klein and object relations, which they suggest is more compatible with discursive psychology in forging a psychosocial methodology. In the same vein, Frosh et al. (2003, p. 40) draw attention to Lacan's 'relentless onslaught on the integrity of the ego' through his account of the mirror stage and the process of misidentification that conceals the inherent lack at the centre of the subject. This sits well within a discursive framework that challenges essentialism and the autonomy of the self. Furthermore, Frosh et al. (2003, p. 40) emphasise the similarities between discourse theory and Lacanian psychoanalysis in their accounts of the social or cultural system as both 'regulatory and productive' of subjectivity where collective discursive resources, or what is referred to as the 'Symbolic' order for Lacanians, both enable expression and limit what can be said. Taken as a way of enhancing existing discursive readings of subjectivity or as a methodological approach in its own right, Lacanian theory has the power to articulate the psychological in a cultural and affective way that ties in with current concerns in tourism studies.

According to the Lacanian thesis, the subject is founded through a traumatic separation from the maternal and entry into the world of signs and language, leaving it forever desiring a lost sense of wholeness. It is this 'lack' that is posited as constitutive of subjectivity and which sets in motion the psychic mechanism of desire. Crucially, Lacan conceptualises lack as fundamental to the subject and desire as something that can never be satisfied. The pivotal role that language, mediation and identification play in this account of subjectivity is to position the essence of the subject beyond its boundaries, as something located in the social matrix beyond the subject rather than a truth buried deep within it. The subject, conceptualised essentially as a void, is thus an 'extimate'[2] one, with that which is experienced as the most personal and intimate being constructed and situated outside itself. Relatedly, Hook (2008) draws attention to Lacan's radical theorisation of the unconscious as a 'trans-individual' phenomenon that can be 'collapsed neither into intrasubjectivity (the unconscious as "Other within me") nor into intersubjectivity (the unconscious as the Other subject)' (see also Chiesa, 2007; Žižek, 1994). This reading of the unconscious departs from more conventional Freudian notions of the unconscious as located within the individual's psyche and thus contributes more readily to an articulation of the psychosocial. This account of the psychological as intersubjective and culturally embedded complicates and resists dualisms of self/other, individual/collective and psyche/social that psychosocial studies have attempted to destabilise, contributing to poststructural thought's move towards more decentred, fragmented accounts of the subject (Wearing & Wearing, 2001). It also provides a way of thinking about extra-discursive dimensions of subjectivity, such as psychic conflicts, desires and other affects in a way that does not reduce the psychological to the individual or cognitive – a concern that has caused some tourism researchers to be hesitant about adopting psychological methods (McCabe, 2005; Moore, 2002).

Through a Lacanian, psychosocial theorisation, we can understand the subject as arising through complex patterns of identification, emotional investments, the play of signs and images, the demands of one's society and irrational impulses arising from the extimate unconscious. By dissolving conventional psychosocial boundaries, a Lacanian ontology allows us to think about how a focus on tourist psychology might simultaneously entail an analysis of the subject's home society and its ideologies. This complements the move towards greater contextualisation of tourists in terms of their everyday lives, cultural context, and the relation between touristic and mundane activities (Franklin & Crang, 2001; Hui, 2008; McCabe, 2002; White & White, 2007). According to a Lacanian lens, to investigate the psychology of a tourist on vacation is simultaneously to study his or her society – the ideologies, injunctions, norms, discourses and affects that permeate the tourist (Kingsbury, 2011, 2005). However, rather than merely implying that the tourist is a passive subject upon which social discourses and norms are inscribed, Lacanian theory provides a way of seeing the social as itself psychically charged through psychosocial mechanisms of fantasy and desire. It is in this way that a Lacanian approach has the potential to foster links between the push for greater contextualisation and the focus on affect and embodiment, because the boundary between self and Other, psychic and social, is theorised as affectively charged, fraught with conflict and lived experientially.

Lacanian tourism studies

Lacanian theory has rarely been used in tourism studies, but those authors who have engaged with it have produced a diverse and thought provoking body of work. Topics explored by these researchers include dark tourism (Buda, 2012; Buda & McIntosh, 2013), volunteer tourism (Crossley, 2012a, 2012b, 2014), the politics of enjoyment in the pleasure periphery (Kingsbury, 2005, 2011), the tourist or host gaze (MacCannell, 2011, 2001; Moufakkir, 2013; Vye, 2004) and theoretical approaches to the unconscious in tourism (Cömert-Agouridas & Agouridas, 2009). Kingsbury (2005, 2011) uses Lacan to explore the ego and fantasy in the context of the Sandals all-inclusive resort in Jamaica. What sets Kingsbury's work apart from other tourism researchers who have delved into psychoanalytic theory is not only the depth of his analysis but also the way in which he fundamentally reconceptualises the spatial ontologies inquiries into tourism are built upon. Kingsbury conceptualises 'psychoanalytic space [as] precariously and terminally liminal, swarming amidst the porosity of borders, spectrality of objects, and the uncanniness of the familiar' (2004, p. 110). In Kingsbury's work, subjectivity is theorised as inherently spatial, using the concept of extimacy to demonstrate the radical ex-centricity of the subject's self and desires. Furthermore, he connects the extimate, intersubjective subjectivities of tourists and resort employees to the broader spatial flows of capital, images and ideologies that frame certain tourism destinations in developing countries as a 'pleasure periphery' (Turner & Ash, 1975).

Kingsbury's account of 'fantasy' in tourism is worth addressing in greater depth and provides a foundation from which my own argument relating to a Lacanian account of cosmopolitan empathy in volunteer tourism is built. Kingsbury (2011, p. 663, original emphasis) argues that in tourism studies '*fantasy* usually refers to tourists' and workers' privately imagined scenarios that stage the fulfilment of desire', yet from a Lacanian perspective 'fantasy provides the very coordinates of tourists' and workers' desire and shields them from the traumatic intrusions of the Real'. Fantasy can thus be conceptualised as the psychic mechanism that teaches us how to desire, providing us with objects of desire rather than merely ways of imagining how we might obtain pre-existing objects of desire. This is one of three crucial differences between a Lacanian and lay understanding of fantasy that Kingsbury delineates. Secondly, in coordinating the objects of our desire, fantasy is also rendered radically intersubjective according to an understanding of desire as belonging to, or being located in, the Other. Thirdly, Kingsbury accentuates the role that fantasy plays in the constitution of reality. Žižek (1997, p. 66) states that 'fantasy is on the side of reality', an observation that makes sense when we take into account the inherence of desire to human subjectivity and fantasy's role in moulding it. To step outside of fantasy would equate to being freed from our perpetual desiring – an ontological impossibility according to Lacan.

Kingsbury uses the Sandals Customer Service Checklist to exemplify how fantasy is orchestrated in a resort setting. The checklist provides employees with tips on how to serve their guests, including prescriptions to always smile, make eye contact, be polite, remember guests' names and introduce oneself when appropriate. Kingsbury argues that this list is not simply a pragmatic set of instructions for ensuring the success of the business, but fantasmatic because 'it instructs workers how to desire … and how to incite guests' desires' (2011, p. 664). These desires are two sides of the same coin given that the workers' desire to appear likeable emerges in response to the tourists' (Other's) desire to be 'serenaded and served by happy Jamaicans' (Kingsbury, 2005, p. 125). Similarly, the Jamaica Tourist Board's 'One Love'[3] mantra can be interpreted as a fantasy that both instructs tourists how to identify the object of their desire in interactions with local people and shields them from 'potentially overwhelming encounters with Jamaicans' enjoyment *qua* the excessive and compulsive choreography of harassment' (Kingsbury, 2005, p. 126). Thus, Kingsbury argues that by addressing libidinal patterns of enjoyment and the operation of ideological fantasies, researchers may be able to articulate critiques of tourism's exploitative practices with a more mature theorisation of subjectivity – an account that brings together the psychological with the political.

As a means of articulating the psychosocial in tourism studies, Lacanian theory is an ideal resource. It allows us to reclaim psychological dimensions integral to tourism's functioning – desire, fantasy, enjoyment, anxiety and shame – in a non-reductive way through the Lacanian conceptualisation of subjectivity as extimate, radically intersubjective and passing endlessly through the Other. The desire of the tourist is experienced on a personal level but emanates from the social field, to which the tourist also belongs and contributes. It is an approach that captures the complexity, fluidity and conflicted nature of subjectivity on both a discursive/symbolic and affective level, possessing the potential to contribute to recent theoretical developments in critical tourism studies that foreground affect and embodiment as essential counterparts to the conventional focus on visual experience (Crossley, 2012a, 2012b; Molz, 2015; Picard & Robinson, 2012; Tucker, 2009). Equally, the heavy emphasis placed on ideology and the political in Lacanian theory lends itself to the branch of enquiry that advocates greater connectivity between our theorisations of tourists in their destinations and their everyday lives at 'home' (Franklin & Crang, 2001; Hui, 2008; McCabe, 2002; White & White, 2007). In the next section, I apply this Lacanian approach to the concept of cosmopolitan empathy and examine how the study of volunteer tourism can be enhanced by adopting a psychosocial perspective on tourist subjectivity.

Towards a psychosocial reading of cosmopolitan empathy

Psychosocial studies provides a way of theorising cosmopolitan empathy as an emergent property of interrelated social and psychic fields. While cosmopolitan empathy already goes beyond generalised empathy in being defined by a certain geographical, social and political context that facilitates affective connection to the Other (Beck, 2006; Höijer, 2004; Kyriakidou, 2008), there is still a prevailing tendency to conceptualise empathy as a private, internal feature of an individual's psychology. Reframing cosmopolitan empathy as inherently psychosocial or, to employ the Lacanian term, 'extimate', invites us to view this affect as at once subjectively experienced and embodied, and as emanating from the social field to which all subjects belong. Although notions of care and empathy may appear antithetical to the alienated Lacanian subject, Lacan does provide insight into the desire to receive or give care and the fantasies underlying these pursuits. Returning to Beck's (2006) conceptualisation of cosmopolitan empathy as the capacity to adopt the perspective of geographically distant others and to feel compassion for their plight, a psychosocial approach invites us to consider this form of empathy not only as connecting subjects imaginatively to 'distant others', but also more fundamentally as an affect that arises intersubjectively and socially.

Conceptualising cosmopolitan empathy as innately social enhances the potential for discourses of global compassion to elucidate the origin and intersubjective dynamics of these affects (Höijer, 2004; Mostafanezhad, 2013a; Sinervo, 2011). Furthermore, if cosmopolitan empathy can be said to be, in a sense, *located* in discourses of global compassion, then the possibility arises that such empathy is ideological as well as discursive and social in nature. This connects strongly to Mostafanezhad's (2014, p. 86) observation that cosmopolitan empathy serves as 'a corollary of the cultural logic and economic policies of neoliberal global capitalism' and sheds light on why the figure of the vulnerable child in developing countries constitutes the predominant object of compassion and empathy in volunteer tourism (Mostafanezhad, 2013a). While it may be the case that volunteer tourists feel for, and aspire to help, children living in poverty in particular parts of the world due to the prominence of such imagery in the media and volunteer tourism marketing, these figures also provide an ideological locus for cosmopolitan empathy to cohere around. The symbol of the vulnerable child permits volunteer tourists to engage with a non-threatening, seemingly apolitical target of development in a way that not only generates the desired emotional experiences seen as critical for personal transformation to occur (Crossley, 2012a, 2012b; Zahra & McIntosh, 2007), but also implicitly maintains a paternalistic power dynamic between active, Northern care-givers and passive, Southern care-receivers (Barnett & Land, 2007; Mostafanezhad, 2013b; Silk, 2004; Sin, 2010). These are discourses and dynamics that serve the interests of the political and economic status quo, and so it is in this sense that cosmopolitan empathy can be conceived of as ideological.

In psychosocial terms, volunteer tourists can be thought of as unconsciously 'invested' in these discourses of the vulnerable, needy Other as constitutive of their identity as a traveller, volunteer or ethical consumer. These representations of the non-threatening, vulnerable Third World child in volunteer tourism resonate with Kingsbury's (2005, p. 126) Lacanian analysis of fantasy in the Jamaican Sandals resort, in which he suggests that fantasy instructs tourists how to identify the object of their desire in interactions with local people and simultaneously shields them from 'potentially overwhelming encounters' such as harassment. Similarly, I argue that there is a fantasmatic quality to the lack presented by these disadvantaged others; they stimulate an affective response – potentially cosmopolitan empathy – that in turn generates desire in the volunteer tourists to enact care and receive gratitude from the visited community. Therefore, by inducing and providing the coordinates for volunteer tourists' desire, this fantasy of the needy Other plays a fundamental role in the constitution of reality in this context (Frosh et al., 2003; Kingsbury, 2011, 2004; Žižek, 1997). The psychic, or fantasmatic, function of cosmopolitan empathy may be in the illusion of connection to the Other that it provides, feeding into the subject's perpetual desire for union and plenitude. This affective response additionally carries the promise of the volunteer tourist being able to enact proximate care towards usually distant others, creating another layer of desired interaction with and recognition from the Other.

This psychosocial investment of volunteer tourists in a fantasmatic, ideological discourse of the vulnerable, needy Other has the potential to challenge conventional spatial understandings of affect and feeling. It is typically assumed that geographical distance presents a barrier to the development of cosmopolitan empathy; that '[d]istance = coldness and nearness = warmth' (Callon & Law, 2004; Mitchell, 2007, p. 706). However, it is possible that distance can have the opposite effect of bolstering volunteer tourists' empathic response to the media images they are exposed to and that closing this distance may result in the disruption of cosmopolitan empathy. In relation to volunteer tourism, there is empirical evidence both to suggest that arrival in the tourism destination can increase the intensity of empathy and related affects through contact with people living in poverty (Zahra, 2011; Zahra & McIntosh, 2007) and can decrease or even eliminate these feelings of concern (Crossley, 2012a, 2012b, 2014). Bridging geographical distance through travel presents volunteer tourists with encounters that can potentially destabilise these discourses underpinning their experience of cosmopolitan empathy. As I have described elsewhere (Crossley, 2012a, 2012b, 2014), volunteer tourists arriving in their travel destination can easily be confronted by local actors who deviate from the dominant image upon which their cosmopolitan empathy is focused. While volunteer tourists expect to meet vulnerable children and adults, eager to receive their help, instead they may encounter locals suspicious of their intentions, precocious youths trying to sell crafts to the tourists or those who seem content in their material poverty rather than in need of help.

The psychosocial analytic concepts of desire, fantasy, investment and disruption bring a unique perspective to understanding cosmopolitan empathy in such a way that can elucidate existing research on the discursive construction of volunteer tourism. I argue that this approach has the potential to show how affects such as cosmopolitan empathy are socially/discursively structured and, equally, how discourse is libidinally/psychically invested. The psychological function that cosmopolitan empathy provides volunteer tourists – in terms of a fantasy of helping poor people in developing countries driven by a desire for recognition from the Other – can arguably be seen as supporting claims that volunteer tourism leads away from an engagement with the politics of development and 'inwardly' towards the volunteer tourist's subjectivity (Butcher & Smith, 2015; Vrasti, 2013). Equally, the potential for fantasy structuring expectations and relationships within the volunteer tourism experience to be disrupted by the spatial transition from empathy at a distance to proximate care demonstrates fragility within the volunteer tourism encounter. This disruption has the potential to shed light on empirical research suggesting that volunteer tourism can actually reinforce cultural stereotypes and 'Othering' rather than engendering cross-cultural understanding and greater global interconnectedness (Palacios, 2010; Simpson, 2004; Snee, 2013; Wearing & Wearing, 2006).

Conclusion

In this paper, I have explored the potential of Lacanian-inflected psychosocial studies to enhance our understanding of cosmopolitan empathy in volunteer tourism, thereby contributing to an emerging critical literature on emotion and affect in tourism (Crossley, 2012a, 2012b, 2014; Molz, 2015; Picard & Robinson, 2012; Tucker, 2016, 2009). This theorisation conceptualises cosmopolitan empathy as an emergent property of interrelated social and psychic fields, which results in the affect serving both ideological and psychological functions. I have argued that bridging geographical distance through travel presents volunteer tourists with encounters that can potentially destabilise the discourses and fantasies of the needy, grateful Other underpinning their experiences of cosmopolitan empathy, thus disrupting the conventional spatial ontology of affect that frequently dominates theoretical discussions of cosmopolitanism. Cosmopolitan empathy is usually linked straightforwardly to the development of global citizenship in volunteer tourism, yet a psychosocial theorisation complicates this relationship by suggesting that the affect may be ideologically driven and open to disruption, rather than a simple internal emotion containing the promise of personal ethical transformation. This paper lays the initial groundwork for further research and, in particular, empirical studies of cosmopolitan empathy and global citizenship in volunteer tourism. Valuable topics for further research could include theoretical work differentiating empathy from related affects, analyses of the relationship between cosmopolitan empathy and motivation in volunteer tourism, and longitudinal studies that explore in greater depth the spatial and affective transition from a state of cosmopolitan empathy to practices of care.

Notes

1. The term 'investment' was first used in this sense by Hollway (1984) to denote the reasons behind a subject's adoption of particular subject positions, replacing 'choice' because this implies a 'rational, decision-making subject' (Hollway & Jefferson, 2005, p. 149).
2. *Extimité* or 'extimacy' is a Lacanian neologism used to suggest that the truth and centre of the subject lies outside itself (Lacan, 2002). The term attaches the prefix *ex* (as in *exterieur*, 'exterior') to the French *intimité* ('intimacy'), thus problematising the boundaries between the inside and outside of the subject.
3. One Love, deriving from the title of Bob Marley's hit song, 'One Love/People Get Ready', has been used in advertisements by the Jamaica Tourist Board since 1994 in an attempt to combat the country's negative reputation for being unsafe and unfriendly.

Disclosure statement

No potential conflict of interest was reported by the author.

Funding

This work was supported by the Economic and Social Research Council (ESRC).

References

Anderson, A. (2001). *The powers of distance: Cosmopolitanism and the cultivation of detachment*. Princeton, NJ: Princeton University Press.

Annette, J. (2002). Service learning in an international context. *Frontiers: The Interdisciplinary Journal of Study Abroad, 8*, 83–93.

Appiah, K. A. (2008). Education for global citizenship. *Yearbook of the National Society for the Study of Education, 107*(1), 83–99.

Appiah, K. A. (2010). *Cosmopolitanism: Ethics in a world of strangers*. New York, NY: Norton.

Baillie Smith, M., & Laurie, N. (2011). International volunteering and development: Global citizenship and neoliberal professionalization today. *Transactions of the Institute of British Geographers, 36*(4), 545–559.

Barnett, C., Cloke, P., Clarke, N., & Malpass, A. (2005). Consuming ethics: Articulating the subjects and spaces of ethical consumption. *Antipode, 37*(1), 23–45.

Barnett, C., Cloke, P., Clarke, N., & Malpass, A. (2010). *Globalizing responsibility: The political rationalities of ethical consumption*. Oxford: Wiley-Blackwell.

Barnett, C., & Land, D. (2007). Geographies of generosity: Beyond the 'moral turn'. *Geoforum, 38*(6), 1065–1075.

Beck, U. (1998). The cosmopolitan manifesto. *New Statesman, 127*(4377), 28–30.

Beck, U. (2006). *The cosmopolitan vision*. Cambridge: Polity Press.

Blackman, L., Cromby, J., Hook, D., Papadopoulos, D., & Walkerdine, V. (2008). Creating subjectivities. *Subjectivity, 22*(1), 1–27.

Brassett, J., & Bulley, D. (2007). Ethics in world politics: Cosmopolitanism and beyond? *International Politics, 44*(1), 1–18.

Brennan, T. (2004). *The transmission of affect*. New York, NY: Cornell University Press.

Brown, S., & Lehto, X. (2005). Travelling with a purpose: Understanding the motives and benefits of volunteer vacationers. *Current Issues in Tourism, 8*(6), 479–496.

Buda, D. M. (2012). *Danger-zone tourism: Emotional performances in Jordan and Palestine* (Doctoral Thesis), University of Waikato, Hamilton, New Zealand.

Buda, D. M., & McIntosh, A. J. (2013). Dark tourism and voyeurism: tourist arrested for "spying" in Iran. *International Journal of Culture, Tourism & Hospitality Research, 7*(3), 214–226.

Butcher, J., & Smith, P. (2010). 'Making a difference': volunteer tourism and development. *Tourism Recreation Research, 35*(1), 27–36.

Butcher, J., & Smith, P. (2015). *Volunteer tourism: The lifestyle politics of international development*. Abingdon: Routledge.

Calhoun, C. J. (2002). The class consciousness of frequent travelers: Toward a critique of actually existing cosmopolitanism. *The South Atlantic Quarterly, 101*(4), 869–897.

Callon, M., & Law, J. (2004). Introduction: Absence-presence, circulation, and encountering in complex space. *Environment & Planning D: Society & Space, 22*(1), 3–11.

Carter, A. (2004). *The political theory of global citizenship*. London: Routledge.

Cheah, P., & Robbins, B. (1998). *Cosmopolitics: Thinking and feeling beyond the nation*. Minneapolis, MN: University of Minnesota Press.

Chiesa, L. (2007). *Subjectivity and otherness: A philosophical reading of Lacan*. Cambridge, MA: MIT Press.

Chouliaraki, L. (2006). *Spectatorship of suffering*. London: Sage.

Clarke, N., Barnett, C., Cloke, P., & Malpass, A. (2007). Globalising the consumer: Doing politics in an ethical register. *Political Geography, 26*(3), 231–249.

Clough, P., & Halley, J. (Eds.). (2007). *The affective turn: Theorizing the social*. London: Duke University Press.

Coghlan, A., & Gooch, M. (2011). Applying a transformative learning framework to volunteer tourism. *Journal of Sustainable Tourism, 19*(6), 713–728.

Cömert-Agouridas, G. B., & Agouridas, D. (2009). The psychoanalytic tourist: Displacement and place of the Other. In J. Parker, L. Tunkrova, & M. Bakari (Eds.), *Metamorphosis and place* (pp. 45–54). Cambridge: Cambridge Scholars.

Crabtree, R. D. (2008). Theoretical foundations for international service-learning. *Michigan Journal of Community Service Learning, 14*, 18–36.

Crossley, É. (2012a). Poor but happy: Volunteer tourists' encounters with poverty. *Tourism Geographies, 14*(2), 235–253.

Crossley, É. (2012b). Affect and moral transformations in young volunteer tourists. In D. Picard & M. Robinson (Eds.), *Emotion in motion: Tourism, affect and transformation* (pp. 85–98). Ashgate: London.

Crossley, É. (2014). *Volunteer tourism, subjectivity and the psychosocial* (Doctoral thesis). Cardiff Univerity, UK. Retrieved from http://orca-mwe.cf.ac.uk/61298/1/FinalPhDthesis.pdf

Delanty, G. (2006). The cosmopolitan imagination: Critical cosmopolitanism and social theory. *The British Journal of Sociology*, 57(1), 25–47.

Derrida, J. (2001). *On cosmopolitanism and forgiveness*. London: Routledge.

Diprose, K. (2012). Critical distance: Doing development education through international volunteering. *Area*, 44, 186–192.

Edley, N. (2006). Never the twain shall meet: A critical appraisal of the combination of discourse and psychoanalytic theory in studies of men and masculinity. *Sex Roles*, 55, 601–608.

Erskine, T. (2002). 'Citizen of nowhere' or 'the point where circles intersect'? Impartialist and embedded cosmopolitanisms. *Review of International Studies*, 28(3), 457–478.

Franklin, A., & Crang, M. (2001). The trouble with tourism and travel theory? *Tourist Studies*, 1(1), 5–22.

Frosh, S. (2001). Things that can't be said: Psychoanalysis and the limits of language. *International Journal of Critical Psychology*, 1, 28–46.

Frosh, S. (2003). Psychosocial studies and psychology: Is a critical approach emerging? *Human Relations*, 56(12), 1545–1567.

Frosh, S., & Baraitser, L. (2008). Psychoanalysis and psychosocial studies. *Psychoanalysis, Culture & Society*, 13(4), 346–365.

Frosh, S., & Emerson, P. (2005). Interpretation and over-interpretation: Disputing the meaning of texts. *Qualitative Research*, 5(3), 307–324.

Frosh, S., Phoenix, A., & Pattman, R. (2003). Taking a stand: Using psychoanalysis to explore the positioning of subjects in discourse. *British Journal of Social Psychology*, 42(1), 39–53.

Frosh, S., & Saville Young, L. (2008). Psychoanalytic approaches to qualitative psychology. In C. Willig & W. Stainton-Rogers (Eds.), *The Sage handbook of qualitative research in psychology* (pp. 109–126). London: Sage.

Giddens, A. (1991). *Modernity and self-identity: Self and society in the late modern age*. Cambridge: Polity Press.

Gilligan, C. (1982). *In a different voice: Psychological theory and women's development*. Cambridge, MA: Harvard University Press.

Gregg, M., & Seigworth, G. (Eds.). (2010). *The affect theory reader*. London: Duke University Press.

Hannerz, U. (2004). Cosmopolitanism. In D. Nugent & J. Vincent (Eds.), *A companion to the anthropology of politics* (pp. 69–85). Oxford: Blackwell.

Hayes, G. (2001). Marxism and psychology: A vignette. *Psychology in Society*, 27, 46–52.

Held, V. (Ed.). (1993). *Feminist morality: Transforming culture, society, and politics*. Chicago: University of Chicago Press.

Held, V. (Ed.). (1995). *Justice and care: Essential readings in feminist ethics*. Boulder, CO: Westview Press.

Henriques, J., Hollway, W., Urwin, C., Venn, C., & Walkerdine, V. (1984). *Changing the subject: Psychology, social regulation and subjectivity*. London: Methuen.

Hepburn, A. (2003). *An introduction to critical social psychology*. London: Sage.

Heron, B. (2011). Challenging indifference to extreme poverty: Southern perspectives on global citizenship and change. *Ethics & Economics*, 8(1), 109–119.

Höijer, B. (2004). The discourse of global compassion: The audience and media reporting of human suffering. *Media, Culture & Society*, 26(4), 513–531.

Hollway, W. (1984). Gender difference and the production of subjectivity. In J. Henriques, W. Hollway, C. Urwin, C. Venn, & V. Walkerdine (Eds.), *Changing the subject: Psychology, social regulation and subjectivity* (pp. 119–152). London: Methuen.

Hollway, W., & Jefferson, T. (2005). Panic and perjury: A psychosocial exploration of agency. *British Journal of Social Psychology*, 44(2), 147–163.

Hook, D. (2005). A critical psychology of the postcolonial. *Theory & Psychology*, 15(4), 475–503.

Hook, D. (2008). Absolute other: Lacan's 'big other' as adjunct to critical social psychological analysis? *Social & Personality Psychology Compass*, 2(1), 51–73.

Hui, A. (2008). Many homes for tourism re-considering spatializations of home and away in tourism mobilities. *Tourist Studies*, 8(3), 291–311.

Jagger, A. (1991). Feminist ethics: Projects, problems, prospects. In C. Card (Ed.), *Feminist ethics* (pp. 78–104). Lawrence: University Press of Kansas.

Jagger, A. (1995). Caring as a feminist practice of moral reason. In V. Held (Ed.), *Justice and care* (pp. 179–202). Boulder, CO: Westview Press.

Jagger, A. M. (1989). Feminist ethics: Some issues for the nineties. *Journal of Social Philosophy*, 20, 91–107.

Kiely, R. (2004). A chameleon with a complex: Searching for transformation in international service-learning. *Michigan Journal of Community Service Learning*, 10(2), 5–20.

Kingsbury, P. (2004). Psychoanalytic approaches. In J. Duncan, N. Johnson, & R. Schein (Eds.), *A companion to cultural geography* (pp. 108–120). Oxford: Blackwell.

Kingsbury, P. (2005). Jamaican tourism and the politics of enjoyment. *Geoforum*, 36(1), 113–132.

Kingsbury, P. (2011). Sociospatial sublimation: The human resources of love in sandals resorts international, Jamaica. *Annals of the Association of American Geographers*, 101(3), 650–669.

Koehn, D. (1998). *Rethinking feminist ethics: Care, trust and empathy*. London: Routledge.

Kyriakidou, M. (2008). 'Feeling the pain of others': Exploring cosmopolitan empathy in relation to distant suffering. In N. Carpentier, P. Pruulman-Vengerfeldt, K. Nordenstreng, M. Harman, P. Vihalemm, B. Cammaerts, H. Nieminen, & T. Olsson (Eds.), *Democracy, journalism and technology: New developments in an enlarged Europe. The Intellectual Work of ECREA's 2008 European media and communication doctoral summer school* (pp. 157–168). Tartu: Tartu University Press.

Lacan, J. (2002). *The seminar of Jacques Lacan: Book XVI: From an Other to the other, 1968-1969* (C.Gallagher, Trans). London: Karnac.

Lawson, V. (2007). Geographies of care and responsibility. *Annals of the Association of American Geographers*, 97(1), 1–11.

Linklater, A. (2007). Distant suffering and cosmopolitan obligations. *International Politics*, 44(1), 19–36.

Lisle, D. (2009). Joyless cosmopolitans: The moral economy of ethical tourism. In M. Paterson & J. Best (Eds.), *Cultural political economy* (pp. 139–157). London: Routledge.

Lyons, K. D., Hanley, J., Wearing, S., & Neil, S. (2012). Gap year volunteer tourism: Myths of global citizenship? *Annals of Tourism Research*, 39(1), 361–378.

Lyons, K. D., & Wearing, S. (2008). All for a good cause? The blurred boundaries of volunteering and tourism. In K. D. Lyons & S. Wearing (Eds.), *Journeys of discovery in volunteer tourism: International case study perspectives* (pp. 147–154). Wallingford: CAB International.

MacCannell, D. (2001). Tourist agency. *Tourist Studies*, 1(1), 23–37.

MacCannell, D. (2011). *The ethics of sightseeing*. Berkeley: University of California Press.

Massey, D. (2004). Geographies of responsibility. *Geografiska Annaler: Series B, Human Geography*, 86(1), 5–18.

McCabe, S. (2002). The tourist experience and everyday life. In G. Dann (Ed.), *The tourist as a metaphor of the social world* (pp. 61–76). Wallingford: CAB International.

McCabe, S. (2005). 'Who is a tourist?' A critical review. *Tourist Studies*, 5(1), 85–106.

McGehee, N. G. (2002). Alternative tourism and social movements. *Annals of Tourism Research*, 29(1), 124–143.

McGehee, N. G., & Norman, W. C. (2001). Alternative tourism as impetus for consciousness-raising. *Tourism Analysis*, 6(3&4), 239–251.

McGehee, N. G., & Santos, C. (2005). Social change, discourse and volunteer tourism. *Annals of Tourism Research*, 32(3), 760–779.

McRobbie, A. (2006). Vulnerability, violence and (cosmopolitan) ethics: Butler's Precarious Life. *The British Journal of Sociology*, 57(1), 69–86.

Mendieta, E. (2009). From imperial to dialogical cosmopolitanism? *Ethics & Global Politics*, 2(3), 241–258.

Mitcheletti, M. (2003). *Political virtue and shopping: Individuals, consumerism and collective action*. London: Palgrave.

Mitchell, K. (2007). Geographies of identity: The intimate cosmopolitan. *Progress in Human Geography*, 31(5), 706–720.

Molz, J. G. (2005). Getting a 'flexible eye': Round-the world travel and scales of cosmopolitan citizenship. *Citizenship Studies*, 9(5), 517–531.

Molz, J. G. (2015). Giving back, doing good, feeling global: The affective flows of family voluntourism. *Journal of Contemporary Ethnography*, 1–27.

Moore, K. (2002). The discursive tourist. In G. Dann (Ed.), *The tourist as a metaphor of the social world* (pp. 41–60). Wallingford: CAB International.

Mostafanezhad, M. (2013a). The geography of compassion in volunteer tourism. *Tourism Geographies*, 15(2), 318–337.

Mostafanezhad, M. (2013b). 'Getting in Touch with your Inner Angelina': Celebrity humanitarianism and the cultural politics of gendered generosity in volunteer tourism. *Third World Quarterly*, 34(3), 485–499.

Mostafanezhad, M. (2014). *Volunteer tourism: Popular humanitarianism in neoliberal times*. Farnham: Ashgate.

Moufakkir, O. (2013). The third gaze: De-constructing the host gaze in the psychoanalysis of tourism. In O. Moufakkir & Y. Reisinger (Eds.), *The host gaze in global tourism* (pp. 203–218). Wallingford: CAB International.

Osler, A., & Starkey, H. (2003). Learning for cosmopolitan citizenship: Theoretical debates and young people's experiences. *Educational Review*, 55(3), 243–254.

Palacios, C. M. (2010). Volunteer tourism, development and education in a postcolonial world: Conceiving global connections beyond aid. *Journal of Sustainable Tourism*, 18(7), 861–878.

Parker, I. (2002). *Critical discursive psychology*. Houndsmills: Palgrave Macmillan.

Parker, I. (2007). *Revolution in psychology*. London: Pluto Press.

Parker, I., & Shotter, J. (Eds.). (1990). *Deconstructing social psychology*. London: Routledge.

Picard, D., & Robinson, M. (Eds.). (2012). *Emotion in motion: Tourism, affect and transformation*. London: Ashgate.

Popke, J. (2006). Geography and ethics: Everyday mediations through care and consumption. *Progress in Human Geography*, 30(4), 504–512.

Popke, J. (2007). Geography and ethics: Spaces of cosmopolitan responsibility. *Progress in Human Geography*, 31(4), 509–518.

Porter, M., & Monard, K. (2001). 'Ayni' in the global village: Building relationships of reciprocity through international service-learning. *Michigan Journal of Community Service Learning*, 8(1), 5–17.

Raymond, E. (2008). 'Make a difference!' The role of sending organizations in volunteer tourism. In K. D. Lyons & S. Wearing (Eds.), *Journeys of discovery in volunteer tourism: International case study perspectives* (pp. 48–60). Wallingford: CAB International.

Raymond, E. M., & Hall, C. M. (2008). The development of cross-cultural (mis)understanding through volunteer tourism. *Journal of Sustainable Tourism*, 16(5), 530–543.

Robinson, F. (1999). The limits of a rights-based approach to international ethics. In T. Evans (Ed.), *Human rights fifty years on: A reappraisal* (pp. 58–76). Oxford: Manchester University Press.

Silk, J. (1998). Caring at a distance. *Philosophy & Geography*, 1(2), 165–182.

Silk, J. (2000). Caring at a distance: (Im)partiality, moral motivation and the ethics of representation. *Ethics, Place & Environment*, 3(3), 303–309.

Silk, J. (2004). Caring at a distance: Gift theory, aid chains and social movements. *Social & Cultural Geography*, 5(2), 229–251.

Simpson, K. (2004). 'Doing development': The gap year, volunteer-tourists and a popular practice of development. *Journal of International Development*, 16, 681–692.

Simpson, K. (2005). Dropping out or signing up? The professionalisation of youth travel. *Antipode*, 37(3), 447–469.

Sin, H. L. (2010). Who are we responsible to? Locals' tales of volunteer tourism. *Geoforum*, 41(6), 983–992.

Sinervo, A. (2011). Connection and disillusion: The moral economy of volunteer tourism in Cusco, Peru. *Childhoods Today*, 5(2), 1–23.

Skrbis, Z., Kendall, G., & Woodward, I. (2004). Locating cosmopolitanism between humanist ideal and grounded social category. *Theory, Culture & Society*, 21(6), 115–136.

Smith, D. M. (1998). How far should we care? On the spatial scope of beneficence. *Progress in Human Geography*, 22, 15–38.

Smith, D. M. (2000). *Moral geographies: Ethics in a world of difference*. Edinburgh: Edinburgh University Press.

Snee, H. (2013). Framing the Other: Cosmopolitanism and the representation of difference in overseas gap year narratives. *The British Journal of Sociology*, 64(1), 142–162.

Standish, A. (2012). *The false promise of global learning: Why education needs boundaries*. London: Continuum.

Swain, M. B. (2009). The cosmopolitan hope of tourism: Critical action and worldmaking vistas. *Tourism Geographies*, 11(4), 505–525.

Swain, M. B. (2016). Embodying cosmopolitan paradigms in tourism research. In A. M. Munar & T. Jamal (Eds.), *Tourism research paradigms: Critical and emergent knowledges* (pp. 87–111). Bingley: Emerald Group Publishing.

Thien, D. (2005). Intimate distances: Considering questions of 'us'. In J. Davidson, L. Bondi, & M. Smith (Eds.), *Emotional geographies* (pp. 191–204). Aldershot: Ashgate.

Tiessen, R., & Epprecht, M. (2012). Introduction: Global citizenship education for learning/volunteering abroad. *Journal of Global Citizenship & Equity Education*, 2(1), 1–12.

Tronto, J. (1993). *Moral boundaries: A political argument for an ethic of care*. New York, NY: Routledge.

Tucker, H. (2009). Recognizing emotion and its postcolonial potentialities: Discomfort and shame in a tourism encounter in Turkey. *Tourism Geographies*, 11(4), 444–461.

Tucker, H. (2016). Empathy and tourism: Limits and possibilities. *Annals of Tourism Research*, 57, 31–43.

Turner, L., & Ash, J. (1975). *The golden hordes: International tourism and the pleasure periphery*. London: Constable.

Vertovec, S., & Cohen, R. (Eds.). (2002). *Conceiving cosmopolitanism: Theory, context and practice*. Oxford: Oxford University Press.

Vrasti, W. (2013). *Volunteer tourism in the global south: Giving back in neoliberal times*. Abingdon: Routledge.

Vye, S. (2004). *Tourist geographies: Spectatorship, space, and empire in England, 1830–1910* (Doctoral thesis). Syracuse University, New York, USA.

Walkerdine, V. (2008). Contextualizing debates about psychosocial studies. *Psychoanalysis, Culture & Society*, 13(4), 341–345.

Wearing, S. (2001). *Volunteer tourism: Experiences that make a difference*. London: CAB International.

Wearing, S., & Neil, J. (2000). Refiguring self and identity through volunteer tourism. *Society and Leisure*, 23(2), 389–419.

Wearing, S., & Wearing, B. (2001). Conceptualizing the selves of tourism. *Leisure Studies*, 20(2), 143–159.

Wearing, S., & Wearing, M. (2006). 'Rereading the subjugating tourist' in neoliberalism: postcolonial otherness and the tourist experience. *Tourism Analysis*, 11(2), 145–162.

White, N. R., & White, P. B. (2007). Home and away: Tourists in a connected world. *Annals of Tourism Research*, 34(1), 88–104.

Zahra, A. (2011). Volunteer tourism as a life-changing experience. In A.M. Benson (Ed.), *Volunteer tourism: Theoretical frameworks and practical applications* (pp. 90–101). Abingdon: Routledge.

Zahra, A., & McIntosh, A. J. (2007). Volunteer tourism: Evidence of cathartic tourist experiences. *Tourism Recreation Research*, 32(1), 115–119.

Žižek, S. (1994). The spectre of ideology. In S. Žižek (Ed.), *Mapping ideology* (pp. 1–33). London: Verso.

Žižek, S. (1997). *The plague of fantasies*. London: Verso.

ⓐ OPEN ACCESS

Class and global citizenship: perspectives from non-elite young people's participation in volunteer tourism

Ruth Cheung Judge

ABSTRACT

Who is 'the global citizen?'. The ideals of global citizenship surrounding volunteer tourism have come under criticism for being invoked in universalising ways, whilst in fact expressing privilege. The assumption in these critiques of the global citizen as 'western, white, middle or upper class, educated, connected' overlooks the diversification of subjects taking part in volunteer tourism, even as it illuminates that we should question the idea of singular, abstract global citizen. This paper – drawing on research with trips run by youth groups based on UK council estates travelling to sub-Saharan Africa – uses perspectives on classed experiences of volunteering to offer some provocations. Firstly, it argues that it is inadequate to invoke a homogeneous figure of the 'privileged volunteer'. Secondly, there is need for more work that asks how contemporary imaginings of 'good works' in the global south are constitutive of subjectivities and exert a political force in the global north, in various ways for particularly positioned subjects. The paper argues that for some subjects, popular ideals of 'global cosmopolitan citizenship' are being drawn into longer-standing projects of reform of the national citizen. However, studying volunteer tourism in practice always reveals ambivalent potential for more emancipatory dynamics and expressions of 'cosmopolitan empathy'.

Introduction: a widespread practice, diverse volunteers

The popularisation of various forms of international volunteering, travel and volunteer tourism to the global south has been remarked upon. Commentators note that such trips have become a widespread rite of passage for many young westerners, arguably one pillar of a renewed 'popular humanitarianism' where charitable work in the global south, particularly in Africa, plays a central role in celebrity culture, corporate marketing and aspirational self-presentation on social media (Daley, 2013; Mathers, 2010; Mostafanezhad, 2013). Short volunteering trips from the UK are facilitated by a range of actors including youth groups, schools, religious groups and diaspora associations. Supported by grants and community fundraising activities, a range of young people beyond those who can afford to pay commercial gap year companies are now participating in volunteer tourism or similar forms of transnational mobility.

This paper is based on research with one such set of initiatives within this diversification – short trips initiated by youth groups based on British council estates, travelling to volunteer in sub-Saharan Africa. The paper uses perspectives on classed experiences of volunteering from this research to offer some provocations to the wider field of scholarship on volunteer tourism as it stands. Firstly, it is inadequate in discussions of volunteer tourism to invoke a homogeneous figure of the 'privileged volunteer'. Secondly, there is need for more work on volunteer tourism that does not just examine and critique the transnational power relations of volunteer tourism in isolation, but asks how contemporary imaginings of 'good works' in the global south are constitutive of subjectivities and exert a political force in the global north (Baillie Smith, 2013), and does so for differently positioned subjects. These provocations draw from – and feed into – theorisations of cosmopolitanism. The volunteer tourists in this research can variously be read as: 'elite cosmopolitans' in terms of enjoying mobility and consuming difference as privileged citizens of western nation-states; 'strategic cosmopolitans' in terms of engaging with the world in ways that respond to the socioeconomic constraints and opportunities they face, and expressing 'cosmopolitan empathy' in terms of making meaningful moves towards respect across difference based on their situated subject

This is an Open Access article distributed under the terms of the Creative Commons Attribution License (http://creativecommons.org/licenses/by/4.0/), which permits unrestricted use, distribution, and reproduction in any medium, provided the original work is properly cited.

positions. Volunteer tourism can be read as produced by – and producing – multiple and ambivalent iterations of 'cosmopolitan global citizenship'.

Volunteer tourism is imagined to 'do good' in two dimensions – in 'helping people' abroad, and in 'bettering' volunteers as individuals. In the cases of this research, the idea of this double benefit was particularly pronounced – the trips studied are imagined to improve and empower 'urban youth' as well as do 'good work' in the volunteering contexts. The words of a participant and a youth worker, below, indicate there are clearly *classed* meanings to these trips:

> I'm the first person to go Africa in my family so to me that's - everyone says like 'oh who's the first one to go uni'. I don't care […] I'll take Africa all day again. (Danny, Volunteer, aged 18)

> They get this sense of worth from [the voluntary work] … which they haven't got here - they feel second class citizens, they feel useless […] they can't get a job, that if they did get a job they wouldn't be able to do it, wouldn't be able to keep it. So it's about working hard, showing they can do it for themselves, about motivation … (Jason, Youth Worker)

Strikingly, Danny explains that participating in a volunteer tourism trip to Zambia is, for him, a marker of pride and achievement on a par with higher education. Participating in 'global charity' is drawn into his aspirations and efforts towards social mobility within the UK. Youth worker Jason believes that the volunteer tourism trips he runs are effective as they provide young people with a sense of achievement and raised aspirations. He hopes that confidence and motivation will have a powerful impact on young peoples' futures, even he half-acknowledges systemic economic constraints. Jason describes his hopes that volunteer tourism – an activity often thought of as epitomising 'global citizenship', doing good outside the borders of one's own nation-state – will improve young people's sense of inclusion and their participation in the UK political-economy, referring to this in terms of their feelings as 'citizens'. Clearly, volunteer tourism has quite particular meanings for these particularly positioned subjects, meanings in which 'global' activities are drawn into identifications and belongings at other levels.

This perspective poses certain provocations. Firstly, it undermines the assumption that the volunteer is always a stand in for 'privilege'. The vast majority of volunteer tourists *are* privileged in terms of global mobility, wealth and their power to extract symbolic resources from travel to the global south. However, exploring heterogeneous positions and experiences *within* this privilege is important – both empirically, in exploring the popularisation of volunteer tourism; and also in terms of interrupting the reification of the association between 'westernness' and privilege to uncover dynamics that might form the basis of more ethical praxis (Griffiths, 2017). Secondly, it begs the question of how ideals of 'global citizenship' are being drawn into building subjectivities for variously positioned volunteers in the global north. Our analysis will be limited if we treat 'global' and 'national' identifications and politics as separate objects. Rather, considering how 'global' citizenships and engagements are always mediated by gendered, classed and racialised hierarchies at localised and national levels (Nagar, Lawson, McDowell, & Hanson, 2002) can provide greater insight into the dynamics that drive, undercut or reshape the power relations involved in various forms of volunteer tourism.

This paper will argue that in the initiatives studied, a charitable ideal of 'global engagement' is being drawn in to longer-standing normative projects to reform young working-class people into 'better' national subjects – though young people participate in volunteer tourism in ways that express their own imaginings of global and national politics. In terms of the possibilities for a politics of cosmopolitan empathy, we see both that certain narratives of global 'helping' are entangled with projects of domination and yet that particular, ambivalent and situated politics of empathy across difference are possible. Cosmopolitanism and global citizenships are multiple and have varied effects in different spheres. Exploring this further, I turn to outline some arguments which push us to consider particularity of global citizenships.

The particularity and multiplicity of volunteer tourism's global citizenships

'*The* global citizen' does not exist. He or she is an abstraction. The idea of 'global citizenship' is always doing something *particular*, despite an appearance of universalism. There are multiple and contradictory practices surrounding the idea and to which the label is attached. Many differentially positioned 'global citizens' have belongings shaped by varied, intersecting identifications. These ideas are not new, but here I outline literature which supports these ideas, and why they are an important basis for thinking about volunteer tourism. Much literature on volunteer tourism has at its heart the debate between whether it might cultivate 'global cosmopolitan citizenship' – conceived as values, attitudes and politics of respect and equality across difference (Jeffrey & McFarlane, 2008; Tiessen, 2011) – or whether it is entrenched in 'neo-colonial' relations, in terms of re-inscribing relationships of dependency and exploitation between former colonial powers and colonies, and the 'global north' and 'global south' more broadly.

On the 'optimistic' side, voices argue that volunteer tourism can be 'an expression and enabler of global citizenship' (Diprose, 2012, p. 5) in terms of tackling cross-cultural stereotypes, building a critical understanding of global inequality and fostering a desire to fight for more equal transnational politics. For instance, Everingham (2015) argues that volunteers at a local organisation promoting language exchange in Ecuador did (though not inevitably) hold critical discourses around development, colonialism and the commodification of volunteer tourism, and experienced feelings of respect, humility and mutuality in the equalising experience of having to speak Spanish. The debate on how to promote such outcomes ranges from emphasising 'well-structured', purposeful volunteering with an explicit emphasis on development education – such as pre-departure training, reflective discussions on social justice, and living and working under local conditions (Diprose, 2012), to suggestions of distancing volunteer tourism from development discourses and emphasising intercultural exchange and solidarity (Palacios, 2010).

On the other side, many strongly conclude that international volunteering is 'neo-colonial' rather than 'globally cosmopolitan'. They see volunteer tourism as based in modernising assumptions of western superiority, and legitimising unskilled young westerners' expertise to 'do development' without attention to skills, sustainability or accountability (Simpson, 2004, p. 683). Critical voices argue that volunteer tourism is reliant on an aestheticisation of poverty as 'authentic' and replaces structural understandings of social justice with depoliticised discourses of individual 'helping'. The focus becomes on volunteer tourists' internal moral transformation, and volunteers' enjoyment is shaped by a sentimental colonial legacy, where needy children epitomise the 'developing world' in need of western help (Crossley, 2012; Darnell, 2011; Mostafanezhad, 2013). Further critiques centre on a reading of volunteer tourism as characteristic of 'neoliberalism': implicated in the broader shift of responsibility for development away from the state to a 'challenge' to be 'tackled' privately and commercially by individuals and corporations (Mostafanezhad, 2013; Simpson, 2005; Sin, Oakes, & Mostafanezhad, 2015). It is read as one facet of 'popular humanitarianism', where 'ethical' products further capital accumulation for multinationals, spectacular events such as Live8 foster consensus that free market capitalism is the 'legitimate' pathway to global justice rather than local activism or alternative transnationalisms (Biccum, 2007), and celebrity work on 'moral issues' depoliticises western interventions (Daley, 2013).

It feels that there is a stalemate between these polarised viewpoints. Perhaps, if we are interested in 'global cosmopolitan citizenships', returning to some of the debates on cosmopolitanism would be helpful. The concept of cosmopolitanism, though subject to definitional debates, is associated with ideals of cross-cultural respect, a capacity to mediate difference, and politics, identities, and practices that cross boundaries, particularly those associated with the nation-state (Jeffrey & McFarlane, 2008; Vertovec & Cohen, 2002). Political scientists' discussions of 'cosmopolitanism' as a sort of universal ethics based on the equal moral worth of individual human beings have been criticised as having a strong European, enlightenment, secular modernist and colonial character (Calhoun, 2002; Gidwani, 2006; Werbner, 1999). The blindness to the particularity of this vision of cosmopolitanism can lead to a tendency to see the west as the site of rationality and progress. Furthermore, existing work can be congratulatory of 'easy forms' of classed consumption: cosmopolitanism as the preserve of those with the political and economic resources to move with ease and consume global cultures (Calhoun, 2002).

Thus, critiques of ideas of cosmopolitanism as a vehicle for 'individualist aspirations and universalist norms' (Breckenridge, Pollock, Bhabha, & Chakrabarty, 2002) which allow 'western societies [to] rewrite their particularity as universalism' (Hall, 2002, p. 28) mirrors critiques of volunteer tourism as a vehicle for 'neo-colonial' and 'neoliberal' imaginaries of 'global citizenship' (Tiessen, 2011). However, where work on volunteer tourism often stops at this critique, critical literature on cosmopolitanism does not dismiss the idea of cosmopolitanism entirely simply because certain visions of it have been fronts for the promotion of values and identities associated with privilege. Instead, they argued that cosmopolitanisms should be understood as plural, situated phenomena, and seek to uncover how people might think and act 'beyond the local' in varied settings (Breckenridge et al., 2002). Explorations of 'grounded', 'vernacular' and 'subaltern' cosmopolitanisms highlight ordinary people's practices of crossing boundaries of difference through 'cultural repertoires' such as labour rights' movements or transnational religious ties (Lamont & Aksartova, 2002). These studies emphasise that 'cosmopolitan' identities and practices can be a strategic resource: for instance, street peddlers in Barcelona share knowledge in cross-cultural networks about how to 'cross borders' spatially (in their migration journeys) and culturally (in business networks, in sales strategies with different nationalities) (Kothari, 2008). That said,

Datta (2009) valuably cautions against making a classed division between elite cosmopolitanism as 'taste' and working-class cosmopolitanism as 'strategic', arguing that London-based Eastern European construction workers' cosmopolitan attitudes arise from *both* 'survival strategies' and an enjoyment and pleasure in engaging with difference. She advocates for more attention to how cosmopolitanisms are shaped by the configurations of power in highly localised spheres of employment, leisure, and domestic space, which produce different interactions.

These critical explorations of cosmopolitanism are valuable for the study of volunteer tourism for several reasons. They help us ask: what type of cosmopolitanism does volunteer tourism foster? An elitist western consumption of difference which becomes fuel for upwards class mobility in the UK? A situated learning of empathy across difference and learning to think beyond the local? Literature on cosmopolitanism problematises any simplistic image of all-powerful global elites, writing that in practice even 'elites' face vulnerability and partiality in their networks and strategies, and still perform 'cosmopolitanisms' shaped by local particularities (Ley, 2004). This does not dismiss a reading of relative privilege, but does problematize a one-dimensional assumption of an archetypal western volunteers. Secondly, despite the fact that the privileged nature of volunteer tourist mobilities remains relevant, global citizenship is not just the preserve of elites. Work on volunteer tourism needs to get beyond this assumption both empirically – since as outlined above, the practice is popularising – and secondly because it reifies that which it seeks to criticise. Literature on cosmopolitanism highlights both that it is important to take a sharp view of power and privilege in relation to how transnational mobilities are determined, articulated and shape identities – particularly avoiding a vague romanticised view of a world of 'flows' (Gogia, 2006), but also to explore the practices and potentials contained in volunteers' engagements across transnational borders which cut against the grain of the problematic power relations that concern us (Griffiths, 2014).

To do so, it is crucial to analyse volunteer tourism in practice. Or put another way, it is crucial to analyse how the multiple possible cosmopolitanisms of volunteer tourism are shaped by the *particular* configurations of power around *particular* forms of volunteer tourism in a diversified field of actors and initiatives. The polarised visions of cynics or apologists for volunteer tourism can lead to 'erasing the social relations through which subjectivities are produced' (Baillie Smith, Laurie, Hopkins, & Olson, 2013, p. 7; Lorimer, 2010; Sin et al., 2015). 'Neo-colonial' dynamics of domination, or 'neoliberal' dynamics of commercialisation and individualisation are not the whole story, and play out in relation to particular national, classed, gendered, racialised and religious identities. For instance, several studies explore how 'global' encounters are shaped by, and shape, the contours of national identity. In Mathers' (2010) study of American travellers (volunteer tourists, political tourists and study abroad students) visiting South Africa (for 1 month to 1 year), the 'reverse gaze' and encounters formed a 'contact zone' where, through the sense of being observed and labelled, travellers faced feelings of discomfort and shock at 'seeing' America for the first time as a despised and adored nation. For many this reinforced a drive to inhabit Americanness by enacting responsibility as citizens of a global power through 'saving Africa'. Others explore the work that 'global encounters' do to national identities in intersection with other identifications. Han (2011) argued that South Korean Christian mission volunteers in East Africa acted and understood their actions through a complex 'assemblage' of selective readings of national history, religious theology and development discourses. Baillie Smith et al. (2013, p. 3) also found that faith-based international volunteering in Latin America fostered hybrid 'global' subjectivities shaped by multiple sources such as 'faith-based imaginaries of global community, public imaginaries of development, discourses around the "gap year"'. Volunteers drew on wider templates in articulation with personal narratives to form a cosmopolitanism which 'smoothed over' understandings of injustice, but also contained moments of critical reflexivity.

This review of literature has aimed to show several things. Firstly, critical literature on volunteer tourism points out that the implicit ideal of 'the global citizen' is: 'western, white, middle or upper class, educated, connected' (Tiessen, 2011). This helps us question the idea of singular, abstract global citizen – but still assumes a certain homogeneity to the subject positions of volunteer tourists. Furthermore, as it stands, literature on volunteer tourism remains rather polarised between such critiques and optimistic voices. Work on cosmopolitanism shows that criticism of elitist, universalising ideologies of cosmopolitanism need not lead to a dead end, but interest in how subjects in varied social positions may express multiple forms of cosmopolitanism which that can be elitist, ambivalent, strategic, empathetic and play out through different spaces and practices (Jeffrey & McFarlane, 2008). This pushes us to move beyond simply repeating totalising critiques of an archetypal elite white volunteer enacting 'global citizenship' as a facade for 'neo-colonialism' and 'neoliberalism', and to ask how the idea of global citizenship interacts with particular volunteers' identifications and strategic navigations of social constraints and opportunities in practice. In other words, wholesale dismissal of volunteer tourism and its visions of 'global citizenship' as always determined by one set of problematic socio-political relations mirrors the universalism it seeks to criticise. Though there might still be much to criticise, it is more interesting to ask how and why the power relations of volunteer tourism play out in particular practices and in relation to particularly positioned subjects. To do so, I now turn to outline the classed projects surrounding the volunteering initiatives in my research.

Global action and classed projects of improvement

This paper is based on research into volunteer tourism trips initiated by council estate based youth groups in London. The youth groups explicitly aimed to work with young people facing socioeconomic vulnerability, and the majority of young participants in the research were framed by youth workers, teachers and local authorities as 'marginalised', 'urban' or 'at risk' in relation to their performance in formal education, residence in low-income households on the estates, proximity to criminal behaviour, and difficult familial situations. Notwithstanding the problematic elision of financial lack with classed and racialised stigma in some of these labels, the overall point is that these trips involved young people positioned very differently to the privileged volunteers whose actions and attitudes are the subject of existing debates. How do volunteer tourism and popular humanitarian imaginaries relate to the subjectivities of those with experiences of exclusion and austerity (Baillie Smith, 2013), who in existing literature are usually framed as being ignored and marginalised by 'spectacles' of global citizenship (Biccum, 2007), the locally-bounded 'other' to the privileged or middle-class 'elite cosmopolitan'? I will first give a little more detail on my data collection before reflecting on how global action and classed citizenships might interact.

I engaged with two main case studies. The first of these was a youth charity, 'Springboard' (n.b. all other organisational and individual names are pseudonyms), based on a council estate in south-west London. As well as work running activity clubs, mentoring programmes and drop-in youth clubs on the estate, overseas trips had been a regular and high-profile strand of the charity's work for nearly 10 years. I accompanied a 10-day trip of theirs to Kenya in February 2013, and interviewed the young participants both before and after the trip. I also undertook weekly participant observation with the charity in London over the course of 14 months, and interviewed youth workers, business funders and young people who had been on trips in past years. The second case study was a youth group from a church, 'Kingsfield Church' based on a council estate in east London. This trip was a one-off initiated by the youth worker. With them, I accompanied a 3-week trip to Zimbabwe in August 2013, and interviewed the young participants both before and after the trip. Beyond the central case studies, I interviewed a number of other key informants and those who had been involved with similar trips. Engaging with young people before, during and after trips provided rich understanding of the way volunteer tourism is framed by collective narratives, *and* plays out through embodied and affective experience, *and* has an extended life through anticipation and memory.

In some ways we might question whether these trips should be defined as 'volunteer tourism'. A much-cited scoping study of volunteer tourism was based on volunteer service organisations (Tourism Research and Marketing [TRAM], 2008), and perhaps small-scale trips initiated by youth groups, schools, religious groups and diaspora associations could be alternatively understood as 'transnational youth work' or 'transnational informal education'. On the other hand, 'volunteer tourism' is a salient label for the trips I studied: they contained a mix of voluntary work and leisure, were short term, and asymmetrical rather than exchange-based. Participants used multiple and shifting framings of the trips, referring interchangeably to 'volunteering', 'mission trips' (in the case of the church youth group) or simply 'holiday' or 'an adventure'. The trips were funded by a mixture of private donations and community fundraising efforts rather than through institutional channels. As in volunteer tourism, they reflected hybrid, popular ideas of doing good which emphasised 'hands-on' helping combined with fun, rather than being tightly defined by state, educational, religious or development-sector visions. Furthermore, the category of 'volunteer tourism' should not be understood as overly coherent, and different initiatives may be shaped to various degrees by ideals such as religiosity, conservation or international development. Therefore, whilst it is worth keeping the distinct characteristics of these trips in mind – as initiatives strongly framed as catalysing 'transitions to adulthood' and fostering particular subjectivities (as explored further in the next section) – I believe the analytical points made from this research are relevant for broader debates on volunteer tourism.

So, definitional disclaimers aside, how can we approach these explicitly classed volunteer tourism initiatives? Broadly, youth travel and volunteering in the global south have been analysed as a practice of middle-class symbolic distinction in the global north, as travel experiences are collected as markers of taste and distinction (Desforges, 1998; Snee, 2013): a mode of gaining cultural capital that fits into a classed 'economy of experience' in the UK where a 'personality package' (a combination of credentials, 'soft skills', and charisma) becomes a marker of employability (Heath, 2007), and the way that a loosely defined 'global consciousness' is valued by contemporary employers (Baillie Smith & Laurie, 2011; Jones, 2011). So we could say that the initiatives under study widen participation in volunteer tourism as a phenomenon in which spatial mobility is central to social mobility. This reading is one where we can see volunteer tourism as a manifestation of 'cosmopolitanism' as a marker of the elite western subject, free to travel, consume and display knowledge of the world (Calhoun, 2002). On the one hand, therefore, we might argue that these initiatives, despite being expressing western privilege, are also equalising opportunities for young people to gain cultural capital within the nation. On the other hand, we can also argue that they are active interventions to 'improve' certain subjects. Take 'Platform2', a national scheme under the UK's last Labour government – a £10 m DfID-funded programme running from 2008 to 2011 – which aimed to widen participation in international volunteering and build development awareness among 'diverse' sectors of the UK population. Platform2's stated aim to '"unlock the potential" within disenfranchised young adults to "become better global citizens"' (DfID 2008 in Diprose, 2012, p. 3) raises intriguing questions about the relation between classed subjectivities and global action.

Why was it seen as important in Platform2 to encourage disenfranchised young people's 'global citizenships'? Diprose (2012, p. 4) raised the criticism that despite some positive outcomes, there is also the sense that such initiatives aim to 'cast[s] disenfranchised young people as grateful, responsible UK citizens'. Indeed, historical work shows that 'global' orientations have long been part of shaping and disciplining the national citizen. In the UK, both formal and informal education has focussed on 'developing' young people in relation to imagined spaces of empire and nation (Collins & Coleman, 2008; Gagen, 2000; Mills, 2013). Youth movements such as the Scouts were deeply shaped by imaginaries of duty to and in the British empire as they made efforts to 'utilise and prioritise the liminal period of youth as a critical and necessary stage in the life course in which to harness and secure an individual's (future) potential and political capital' (Mills, 2013, p. 123). But more specifically, within Scouting's efforts to shape young Britons' characters, working-class youth were seen as particularly in need of scouting's messages of responsibility, duty and self-regulation (Mills, 2013). Moral education within the nation and imaginings of abroad have long been linked. I turn to explore how this played out in my research now.

Global citizenship and disciplining the national citizen

Encounters with global 'others' achieve work at the interior frontiers of citizenship – such as the boundaries of class – as well as at the outer boundaries of nation (Stoler, 2001). This was highly evident in the cases of this research. In the discourses surrounding the trips, concerns with reforming non-elite young people conceived as 'at risk' or problematic were fused with contemporary popular humanitarian ideals of 'saving the world'. This is evident in the words, below, of a business funder of one of the youth charity, whose idea that the trips help 'change' people' into being 'good citizens' implies that young participants are currently not. These ideas were also to a certain extent internalised by the young participants. Dylan, in a pre-trip interview, expressed the idea that change in his group of friends (referred to at a distance – 'kids') would be driven by gaining 'perspective' about comparative privilege:

> The people that [Springboard, the youth charity] target are the people that are ready to change their life … […] hopefully … they kind of flick over into being a good citizen, so to speak. (Hamish, Springboard Business Funder)

> … the thing that Springboard are doing makes Roehampton a better place – like, taking loads of kids from Roehampton to like, see … places like Kenya … Everyone's going to have a different mindset when we come back. […] I think the way they live over there is going to be so bad. Like, there's just pure poverty over there. (Dylan, Volunteer, aged 16)

In the initiatives this research engaged with, participants saw themselves as becoming transformed subjects: more 'grateful', 'charitable' and 'motivated'. This narrative of personal moral transformation, also observed in work on volunteer tourism with middle-class participants (Crossley, 2012; Darnell, 2011), takes on a different character in relation to differently positioned subjects. There is a heightened disciplinary aspect to the way imaginaries of global charity mesh with longer-standing efforts at reforming young working-class people who are especially subject to heavily moralised inscriptions of adult hopes and fears (Kraftl, 2008; Valentine, 1996).

The normalisation of the idea of a short trip abroad as leading to transformation at home is underpinned by powerful spatial imaginaries of 'here' and 'there'. The belief in the power of the trip rests on the heavy coding of two spatial contexts – urban London and sub-Saharan Africa – the first pathologised as violent, the second objectified as a space of victimhood and, as expressed Dylan's words, 'pure poverty'. Ideals of improving young subjects are expressed through the assumption that the global south is also in need of improvement (Aitken, Lund, & Kjørholt, 2007). My research strongly confirmed the power of 'neo-colonial' imaginaries in volunteer tourism (Mostafanezhad, 2013) – destinations framed as depersonalised needy spaces, and strong imaginaries of 'Africa' as an object of care, as the locus of shocking poverty, inherent virtue or challenging risk, with an almost complete lack of interest in the history, politics or simply factual knowledge of African nations. The key point here is that these problematic visions were not just evidence of 'western privilege' but also a core underpinning of disciplining particular young subjects within the boundaries of a privileged nation. We might ask then what are the different conditions under which a 'cosmopolitan' interest with the world is framed, and the socio-political milieus that make adopting and performing a certain version of 'cosmopolitanism' make strategic sense for young people.

The most extreme visions of volunteer tourism trips as reforming apathetic, lazy or potentially criminal youth through hard work and a sense of comparative privilege within my research were expressed by Springboard's business funders, wealthy individuals who lived locally to the estate and provided generous and untied financial support, and sometimes more hands-on help – for the youth charity's work. Stephanie, below, talks about her idea of the trips as a sort of 'shock therapy' to make young people realise their comparative privilege and prompting them to work hard, beginning with fundraising efforts and continuing on the trips:

> They ... got to see what poverty was really all about, and actually that their lives weren't so bad ... yes, they are in the lower, you know, demographic, but [...] these guys have a roof over their head, they've got a free education, free NHS ... [...] Obviously they were made to fundraise in advance ... although ... I still feel they should be pushed a bit harder. (Stephanie, Springboard Business Funder)

Universalising ideas of Kenya as defined by need and poverty underpin the idea that it is a good backdrop against which to use volunteer tourism as a chance to gain 'perspective' and practice hard work – in other words, to shape ungrateful youth in low-income Britain into the 'deserving poor' who work hard, do not cause problems or see themselves as victims, and are not going to become 'dependent' or feel entitled. It should be noted that these views were not representative of Springboard youth workers, who had much more generous and holistic views of the young people they worked with. However, these views are worth quoting because they illustrate the logical conclusions of many of the implicit assumptions present in other accounts – including of youth workers and young people – that it is good that the trips help push young people to exert individual efforts at social mobility. For instance, Jacob expresses this idea:

> It kind of helped me realise – I need to step up a bit [...] No-one's going to make you ... get up and do stuff. You've got to get up and make yourself do stuff. (Jacob, Volunteer, aged 17)

The invocation to self-discipline and motivation contained in these visions of the 'good' volunteer tourism achieves for the participant is problematic as it de-legitimises young people's very real struggles by slotting in neatly with a politics of 'responsibilisation of poverty': the idea that working-class individuals should taking responsibility for their own betterment amid the dismantling of the welfare state (Allen, Hollingworth, Mansaray, & Taylor, 2013).

Thus, in these particularly positioned cases, becoming a 'global citizen' is part of a pressure to become an 'aspirational' citizen within the nation–or put differently, a pressure to adopt a sort of 'strategic cosmopolitanism' shaped by the constraints of classed prejudice. A 'politics of aspiration' has been identified as amplifying disciplinary pressures on working-class and racialised young people to disavow disparaged ways of being an 'adjust themselves' to succeed (Allen et al., 2013; Brown, 2013; Kulz, 2014). Brown (2013) examines how discourses of aspiration education policies focus interventions on working-class young people deemed 'not aspirational enough'. Young people's cultivated ambitions for higher education come into painful conflict with facing the difficulties of the withdrawal of state support (e.g. high university fees) and actually achieving social mobility post-university. An emphasis on aspiration as the driver of social mobility ends up 'locat[ing] the "blame" for disadvantage or inequalities in the outcomes of young people's lives within the (pathologised) working class/minority individual' (Archer, Halsall, & Hollingworth, 2007, p. 562).

In this research, volunteer tourism was drawn into quite intimate re-formations of the self as aspirational. For instance, Richie, below, describes volunteer tourism having prompted him to try to adjust his dress, distancing himself from conspicuous consumption as intersecting with the denigrating label of 'gangster' overlaid on him as a young black man, and communicating 'motivation':

> Before the trip ... I'd think 'ah I really NEED a new pair of trainers [...] but I realise that's not really doing me anything [...] Cos I don't wanna seem as – uh ... 'gangster' ... I've never dressed like a gangster, but I'm just saying like – I wanna have a new look to me, I wanna look more smart, I wanna look more motivated ...
> (Richie, Volunteer, aged 16)

The ideal that volunteer tourism might foster aspirational and enterprising dispositions, was strongly pronounced in high-profile 'success stories' of ex-drug dealers turned young entrepreneurs – a transformation believed to have been catalysed by volunteer tourism as providing a sort of 'affective energy' of 'adventure' and efficacy ('making a difference'). In these trips, explicit ideas of 'global citizenship' were quite vague, but in practice, relations of charitable care towards infantilised others and active manual labour read as modernising 'development' were central to imaginings of these projects of transforming volunteers into a grateful, hardworking subjects. Other ideas of 'good works' were conceived of as supporting aspiration of those we met, wanting to support those in the destination to sell their wares or use their talents of self-presentation and bodily skill to become entertainers or sports people. These visions were less objectifying of those in the destination, but clearly emerged out of a resonance around aspiration, enterprise and individual responsibility in what it means to make part in global action and the type of economic subjects upheld as ideal.

Furthermore, the aspirational subjectivities that young people returned from trips with included global charity. These were visions of individual capitalist success with a philanthropic edge informed by the 'celebrity-corporate-charity complex' (Brockington, 2014). For instance, Jamie voices dreams of being charitable meshed with dreams of sporting success, where poverty alleviation is achieved through the generous giving of resources from the massive personal wealth of some individuals to other, poor individuals, rather than in strengthening social safety nets, or systems of redistribution:

> My dream would be to be a footballer. ... partly because the amount of money I'd have ... like I could go to Zimbabwe and literally – like, literally just like that – they'd have all the money they need ... (Jamie, Volunteer, aged 15)

Thus, the findings from this research can be analysed in line with approaches to volunteer tourism as a 'technology of the self' 'through which subjects constitute themselves simultaneously as competitive, entrepreneurial, market-based, individualised actors and caring, responsible, active, global citizens' (Sin et al., 2015, p. 122).

However, this is not to advocate for readings of volunteer tourism as always or merely 'neoliberal governmentality' in a way that applies a stop-gap analysis that tells us little. Rather, the point is that in this particular setting, the popularisation of doing good abroad has been drawn into pressures on young subjects to take individual responsibility for their own betterment in contemporary Britain. The criticism that comes into stark relief when studying initiatives targeted at working-class youth is that vague ideals of global citizenship are actually preparing young people to be certain types of national citizens, amenable and self-managing amid an insecure and flexible labour market, through encouraging 'enterprising' (overcoming limits, taking opportunities, realising ambitions) subjectivities (Cremin, 2007; Gagen, 2015).

This section has argued that in the initiatives studied, 'global citizenship' is enrolled in a project of classed reform within the nation. This reveals the strong resonances between the neo-colonial expressions of global citizenship often (not inevitably) contained in volunteer tourism and the way certain subjects within the nation are framed as in need of civilisation. Gagen (2007) explores how in early twentieth century USA there was an 'interpenetration' of stories about racialised backwardness in America's imperial territories and America's urban poor children. Casting colonised people as child-like was mirrored by ideas that *all* children were 'primitive' and in need of being 'civilised', underpinning a logic of improvement enacted both overseas and in urban reformers' work to create modern civilised citizens. The trajectory of the improvement in these cases of volunteer tourism were framed within classed ideals of an elite cosmopolitan subjectivity that is very much embedded in western-capitalist engagements global difference. However, young participants did not only adopt the self-disciplining aspirational subjectivities through the volunteer tourism trips, but also made their own meanings around global and national citizenship, and it is to this I turn now.

Non-elite cosmopolitanisms and ongoing potentials of global citizenship

Overall, as demonstrated above, volunteers adopted stories of reform through volunteer tourism into aspirational subjects, speaking to the lack of 'thinkable spaces' for creating a valued self and future amid classed and racialised hierarchies – such as Richie's need to 'adjust himself' to contend with the 'gangster' label. These adjustments can be read as an agentive expressions of determination to survive and thrive amid the withdrawal of social safety nets (Katz, 2001), even as they 'replenish established power structures' (Jeffrey, 2012, p. 249). However, there were other stories at play, and young people used the experience of volunteer tourism for their own ends, constructing their own meanings (Wood, 2012). In this section I will briefly outline two ways they did so: firstly, by relating to those they met in the destination contexts in ways which contained elements of class-based solidarity, and secondly, by upholding participation in volunteer tourism as proof of their existing virtue rather than their need for reform. These dynamics can be read as testifying to the enduring potential of volunteer tourism to contain potentials for 'cosmopolitan empathy'.

Despite the strength of framings of volunteer tourism described above – as practices of charitable pity and modernising western improvement – young volunteers did express connections with destinations and people within them which ran counter to relations of charitable pity. These were multifarious. Young people engaged in friendly exchanges through ordinary affinities of gender, age and personality, bonding over playing games, football, affectionate insult-trading and gendered humour. Young people with second-generation African heritage expressed claims of special and proud connection to Africa (see Cheung Judge, 2016). Christian religiosity provided a platform for those in the destination context to assert equality and even superiority in ways which undercut dynamics of western charitable virtue. Transnational youth culture – such as pop music, and 'urban' style – provided moments of pleasurable connection which blurred the boundaries of global and local (see Cheung Judge, 2017).

However, perhaps the most interesting way young people expressed desires to connect across transnational boundaries which went against the grain of 'neo-colonial' dynamics in volunteer tourism were connections centred on resonances of economic insecurity and intersectional prejudice, and a spirit of resistance in the face of this. For instance, one young volunteer, whose experience of the educational system had been deeply exclusionary, spoke of 'identifying' with a child who was reputedly 'the bad kid of the school' (Research Diary, Kenya Trip). In Kenya, we volunteered at a home for street children, and I listened to a group of young men from south London talk with awed admiration to one of the older boys there about being chased by the police, street violence and stealing to make a living. In these moments there was an ambivalent mixture of mutual connection and voyeurism (Research Diary, Kenya Trip). On the one hand, there was a bonding over relatable experiences of police aggression, illegal ways of making a living and strong ties of loyalty: a transnational resonance around urban poverty. However, there were also elements of a fascinated consumption of the young man's life story which contained a sort of appropriation of it as drama on demand rather than as something recognised as having a gravity and particularity of its own which required respect.

However, despite these blurred lines, such connections represent a profound potential for thinking volunteer tourism trips differently: they contain the idea that economic inequality and suffering is not the natural preserve of some regions of the world and not others, and suggest 'helping' could be reimagined as based on solidarity around the relational impacts of inequality. Griffiths (2017) argues for more attention to social class to expose the heterogeneity of north–south relations, in that classed experiences can sensitise us to the pain of unequal opportunities, being assigned social positions, and experiences of exclusion. Whilst relative privilege must be always acknowledged, such resonances may form the basis of speaking *to* not *for*, others. This echoes work by feminist scholars on cosmopolitanism and ethics, which argues that mutual recognition of bodily vulnerabilities, loss and pain may be a basis for an ethics of care which crosses transnational borders (McRobbie, 2006; Mitchell, 2007). In classed experiences of volunteer tourism, we might wonder whether there is scope for young subject to draw on a shared experience of stigmatisation around poverty.

These are only fragments of solidarity, but they do highlight that cosmopolitan global citizenships are not exclusive with other identifications, but rather particular identities – whether religious, gendered, aged or stemming from particular 'urban' contexts – can act as resources in supporting connection to global 'others' (Calhoun, 2002). Furthermore, we can see the performance of cosmopolitanism by those in the destination contexts. For instance, the head of the children's home in Kenya, Joseph, had been a street child in his youth. He told his life story consciously using language to link to the UK-context from which the UK group came: describing himself as from a 'Kenyan council estate', in 'gangs', and 'doing drugs'. Joseph's cosmopolitan mobilisation of vocabularies of street 'realness' was an exertion of authority to actively participate in the 'reform' of the young UK subjects.

These observations also highlight that 'friendship politics' are often the mode through which young volunteers express agentive intentions or attempts to connect across difference (Wood, 2012). I overhead one young man say to a Kenyan young person he met that 'the ghetto is the most welcoming place' and 'my house isn't big, but you could always stay' (Research Diary, Kenya Trip). This statement contains multiple implicit critiques: of classed snobbery, of the elitist forms of cosmopolitanism, and of the lack of genuine exchange that underpins volunteer tourist flows. Although these moments should not be read as a deeply realigned understanding of global relations (Diprose, 2012), they may be a start.

So, if young people reworked the charitable ideals of 'global citizenship' in the trips through friendship politics, they also engaged with the process in ways which reworked the idea of reforming into better national citizens. Young volunteers talked about how going on the trips gave them a sense of worth, in their own eyes and those of others. Rather than the young people expressing that trips transformed them from 'bad' to 'good' subjects, they celebrated the trips as 'proof' of their pre-existing worth, virtue and abilities in the face of classed and racialised (as intersecting with gendered) prejudice. These expressions were more than just assertions of aspirational subjectivities, rather about using the trips to counteract being seen as defined by classed 'lack'. For Lisa, below, memories become material towards 'seeing herself' as virtuous against internalised judgements of worth based on capital accumulation:

> It makes you look at yourself in a different light, like I'm – not a bad person … makes you feel better in yourself […] older people especially, like, upper class people – can look down on you […] yeah, I don't earn a lot of money, but I've been to Africa, and I've helped people … (Lisa, Volunter, aged 19)

Clearly, these feelings remain underpinned by ideas of 'helping others' and self-improvement. However, quotes like that of Lisa help us read young people's desires to 'change' not as evidence they fully submit to disciplinary forces that cast them as criminal or apathetic – but as a way to navigate their way blockages to social mobility and of prejudice: volunteer tourism trips being engaged in search of respect.

Conclusion

This paper has shared perspectives from some quite heavily classed experiences of volunteer tourism: initiatives which target 'urban youth'. From this, it has offered some provocations which may be of interest to the wider field of thinking on volunteer tourism. The first of these is to call attention to the diversity of subjects and initiatives engaging in short term, popular forms of mobility from the global north to the global south. Exploring these mobilities in practice reflects the value in interrogating an undifferentiated category of 'western privilege'. Findings from these volunteer tourism initiatives expose the way that for some subjects, popular ideals of global citizenship are drawn into longer-standing projects of reform of the national citizen. In particular, 'aspirational' performances of citizenship which include aspirations towards a criticized 'elitist, western' version of cosmopolitanism are being inculcated and negotiated through practices of volunteer tourism.

Secondly, these findings point out that particular articulations of 'cosmopolitan global citizenship' do work in different social spheres, including within the nation. Whilst this observation does not directly contradict analyses of 'global citizenship' as a story of formal politics as replaced by consumption-based interactions, it does assert that volunteer tourism is being engaged in struggles over collective politics with a small 'p' within the nation. Volunteer tourism might contain and promote de-politicised understandings of how to achieve social change, but it is still entangled in the politics of social change itself as a practice. The ways that these initiatives are normatively framed as reforming potentially deviant or apathetic youth is one example of the ways that young people are particular objects of adult societies' political-economic concerns (Jeffrey, 2012; Katz, 2008), and asked to become subjects that accede to a flexible, insecure economy. Here, young people's embrace of volunteer tourism can be read in the light of the way that cosmopolitan subjectivities are often a 'strategic' response to socioeconomic contexts and framings. Yet the ways that young people's own intentional acts of empathy across difference or to articulate their own meanings around participating in volunteer tourism can also be read as responses to the political-economic pressures and the social contexts they navigate (Skelton, 2013).

Thirdly, the multiple and contradictory dynamics exposed by studying volunteer tourism in practice points us towards seeing that any potential for 'cosmopolitanism', as an equalising ethics of relating across difference, will always be enfolded in particular social dynamics and may be ambivalent (Jeffrey & McFarlane, 2008). Where some work on volunteer tourism can fall into rather black-and-white polemical statements about whether volunteer tourism is 'all good' or 'all bad', arguably:

> Asking if international volunteering creates cosmopolitans or global citizens provides too rigid and instrumental an approach, focusing attention on a particular status or end point and erasing the social relations through which subjectivities are produced … [this] risks excluding the multiple languages in which global or cosmopolitan subjectivities may be expressed. (Baillie Smith et al., 2013, p. 7)

Perhaps we might better see the 'cosmopolitan' efforts of volunteer tourism as ambivalent, dependent on particular practices: at times an expression of western privilege to consume difference, at times a 'technology of rule' creating self-regulating subjects who have responsibilities beyond but simultaneously to the nation, and at times containing a currency and potential to encourage lived practices of ethical acts of care, respect and tolerance that are deeply necessary given the many 'structures of violence' in society (Mitchell, 2007). The cases recounted here bring up sharp questions around class – volunteer tourism must be viewed critically where it reinforces classed inequalities, and yet there is potential in widening class participation in volunteer tourism for transnational solidarity and the recognition of the worth and virtue of young subjects who face classed prejudice. Future scholarship must analyse volunteer tourism's potential for 'fostering global citizenship' in the light of how multiple cosmopolitanisms – from the dominant, to the strategic, to the empathetic – are shaped by the particular configurations of power in diverse volunteer tourism initiatives.

Disclosure statement

No potential conflict of interest was reported by the author.

Funding

This research was conducted during a +3 PhD Studentship supported by the Economic and Social Research Council.

References

Aitken, S. C., Lund, R., & Kjørholt, A. T. (2007). Why children? Why now? *Children's Geographies*, 5(1–2), 3–14.

Allen, K., Hollingworth, S., Mansaray, A., & Taylor, Y. (2013). Collisions, coalitions and riotous subjects: Reflections, repercussions and reverberations – an introduction. *Sociological Research Online*, 18(4). doi:10.5153/sro.3149

Archer, L., Halsall, A., & Hollingworth, S. (2007). Inner-city femininities and education: "Race", class, gender and schooling in young women's lives. *Gender and Education*, 19(5), 549–568.

Baillie Smith, M. (2013). Public imaginaries of development and complex subjectivities: The challenge for development studies. *Canadian Journal of Development Studies/Revue Canadienne D'études Du Développement*, 34(3), 400–415.

Baillie Smith, M., & Laurie, N. (2011). International volunteering and development: Global citizenship and neoliberal professionalisation today. *Transactions of the Institute of British Geographers*, 36(4), 545–559.

Baillie Smith, M., Laurie, N., Hopkins, P., & Olson, E. (2013). International volunteering, faith and subjectivity: Negotiating cosmopolitanism, citizenship and development. *Geoforum*, 45, 126–135.

Biccum, A. (2007). Marketing development: Live 8 and the production of the global citizen. *Development and Change*, 38(6), 1111–1126.

Breckenridge, C. A., Pollock, S., Bhabha, H. K., & Chakrabarty, D. (Eds.). (2002). *Cosmopolitanism*. Durham, NC: Duke University Press Books.

Brockington, D. (2014). The production and construction of celebrity advocacy in international development. *Third World Quarterly*, 35(1), 88–108.

Brown, G. (2013). The revolt of aspirations: Contesting neoliberal social hope. *ACME: An International E-Journal for Critical Geographies*, 12(3), 419–430.

Calhoun, C. (2002). The class consciousness of frequent travellers: Towards a critique of actually existing cosmopolitanism. In S. Vertovec & R. Cohen (Eds.), *Conceiving cosmopolitanism: Theory, context, and practice* (pp. 86–109). Oxford: Oxford University Press.

Cheung Judge, R. (2016). Negotiating blackness: Young British volunteers' embodied performances of race as they travel from Hackney to Zimbabwe. *Young*, 24(3), 238–254.

Cheung Judge, R. (2017). Volunteer tourism and nonelite young subjects: Local, global, and situated. In C. Dwyer & N. Worth (Eds.), *Geographies of identities and subjectivities volume 4 in Skelton, T. (editor-in-chief) geographies of children and young people* (pp. 249–268). Singapore: Springer.

Collins, D., & Coleman, T. (2008). Social geographies of education: Looking within, and beyond, school boundaries. *Geography Compass*, 2(1), 281–299.

Cremin, C. (2007). Living and really living: The gap year and the commodification of the continent. *Ephemera*, 7(4), 526–542.

Crossley, É. (2012). Poor but happy: Volunteer tourists' encounters with poverty. *Tourism Geographies*, 14(2), 235–253.

Daley, P. (2013). Rescuing African bodies: Celebrities, consumerism and neoliberal humanitarianism. *Review of African Political Economy*, 40(137), 375–393.

Darnell, S. C. (2011). Identity and learning in international volunteerism: "Sport for development and peace" internships. *Development in Practice*, 21(7), 974–986.

Datta, A. (2009). Places of everyday cosmopolitanisms: East European construction workers in London. *Environment and Planning A*, 41(2), 353–370.

Desforges, L. (1998). "Checking out the planet": Global representations/local identities and youth travel. In T. Skelton & G. Valentine (Eds.), *Cool places: Geographies of youth cultures* (pp. 175–192). New York, NY: Routledge.

Diprose, K. (2012). Critical distance: Doing development education through international volunteering. *Area*, 44(2), 186–192.

Everingham, P. (2015). Intercultural exchange and mutuality in volunteer tourism: The case of intercambio in Ecuador. *Tourist Studies*, 15(2), 175–190.

Gagen, E. A. (2000). An example to us all: Child development and identity construction in early 20th-century playgrounds. *Environment and Planning A*, 32(4), 599–616.

Gagen, E. A. (2007). Reflections of primitivism: Development, progress and civilization in imperial America, 1898–1914. *Children's Geographies*, 5(1–2), 15–28.

Gagen, E. A. (2015). Governing emotions: Citizenship, neuroscience and the education of youth. *Transactions of the Institute of British Geographers, 40*(1), 140–152.

Gidwani, V. K. (2006). Subaltern cosmopolitanism as politics. *Antipode, 38*(1), 7–21.

Gogia, N. (2006). Unpacking corporeal mobilities: The global voyages of labour and leisure. *Environment and Planning A, 38*(2), 359–375.

Griffiths, M. (2014). The affective spaces of global civil society and why they matter. *Emotion, Space and Society, 11*, 89–95.

Griffiths, M. (2017). From heterogeneous worlds: Western privilege, class and positionality in the South. *Area, 49*(1), 2–8.

Hall, S. (2002). Political belonging in a world of multiple identities. In S. Vertovec & R. Cohen (Eds), *Conceiving cosmopolitanism: Theory, context, and practice* (pp. 25–31). Oxford: Oxford University Press.

Han, J. H. J. (2011). 'If you don't work, you don't eat': Evangelizing development in Africa. In J. Song (Ed.), *New millennium South Korea: Neoliberal capital and transnational movements* (pp. 142–158). New York, NY: Routledge.

Heath, S. (2007). Widening the gap: Pre-university gap years and the "economy of experience". *British Journal of Sociology of Education, 28*(1), 89–103.

Jeffrey, C. (2012). Geographies of children and youth II Global youth agency. *Progress in Human Geography, 36*(2), 245–253.

Jeffrey, C., & McFarlane, C. (2008). Performing cosmopolitanism. *Environment and Planning D: Society and Space, 26*, 420–427.

Jones, A. (2011). Theorising international youth volunteering: Training for global (corporate) work? *Transactions of the Institute of British Geographers, 36*(4), 530–544.

Katz, C. (2001). Vagabond capitalism and the necessity of social reproduction. *Antipode, 33*(4), 709–728.

Katz, C. (2008). Cultural geographies lecture: Childhood as spectacle: Relays of anxiety and the reconfiguration of the child. *Cultural Geographies, 15*(1), 5–17.

Kothari, U. (2008). Global peddlers and local networks: Migrant cosmopolitanisms. *Environment and Planning D: Society and Space, 26*(3), 500–516.

Kraftl, P. (2008). Young people, hope, and childhood-hope. *Space and Culture, 11*(2), 81–92.

Kulz, C. (2014). "Structure liberates?": Mixing for mobility and the cultural transformation of "urban children" in a London academy. *Ethnic and Racial Studies, 37*(4), 685–701.

Lamont, M., & Aksartova, S. (2002). Ordinary cosmopolitanisms strategies for bridging racial boundaries among working-class Men. *Theory, Culture & Society, 19*(4), 1–25.

Ley, D. (2004). Transnational spaces and everyday lives. *Transactions of the Institute of British Geographers, 29*(2), 151–164.

Lorimer, J. (2010). International conservation "volunteering" and the geographies of global environmental citizenship. *Political Geography, 29*(6), 311–322.

Mathers, K. (2010). *Travel, humanitarianism, and becoming American in Africa*. New York: Palgrave Macmillan US.

McRobbie, A. (2006). Vulnerability, violence and (cosmopolitan) ethics: Butler's Precarious Life. *The British Journal of Sociology, 57*(1), 69–86.

Mills, S. (2013). 'An instruction in good citizenship': Scouting and the historical geographies of citizenship education. *Transactions of the Institute of British Geographers, 38*(1), 120–134.

Mitchell, K. (2007). Geographies of identity: The intimate cosmopolitan. *Progress in Human Geography, 31*(5), 706–720.

Mostafanezhad, M. (2013). "Getting in touch with your inner Angelina": Celebrity humanitarianism and the cultural politics of gendered generosity in volunteer tourism. *Third World Quarterly, 34*(3), 485–499.

Nagar, R., Lawson, V., McDowell, L., & Hanson, S. (2002). Locating globalization: Feminist (Re)readings of the subjects and spaces of globalization. *Economic Geography, 78*(3), 257–284.

Palacios, C. M. (2010). Volunteer tourism, development and education in a postcolonial world: Conceiving global connections beyond aid. *Journal of Sustainable Tourism, 18*(7), 861–878.

Simpson, K. (2004). "Doing development": the gap year, volunteer-tourists and a popular practice of development. *Journal of International Development, 16*(5), 681–692.

Simpson, K. (2005). Dropping Out or signing Up? The professionalisation of youth travel. *Antipode, 37*(3), 447–469.

Sin, H. L., Oakes, T., & Mostafanezhad, M. (2015). Traveling for a cause: Critical examinations of volunteer tourism and social justice. *Tourist Studies, 15*(2), 119–131.

Skelton, T. (2013). Young people, children, politics and space: A decade of youthful political geography scholarship 2003–13. *Space and Polity, 17*(1), 123–136.

Snee, H. (2013). Framing the other: Cosmopolitanism and the representation of difference in overseas gap year narratives. *The British Journal of Sociology, 64*(1), 142–162.

Stoler, A. L. (2001). Tense and tender ties: The politics of comparison in North American history and (post) colonial studies. *The Journal of American History, 88*(3), 829–865.

Tiessen, R. (2011). Global subjects or objects of globalisation? The promotion of global citizenship in organisations offering sport for development and/or peace programmes. *Third World Quarterly, 32*(3), 571–587.

Tourism Research and Marketing. (2008). *Volunteer tourism: A global analysis: A report*. ATLAS.

Valentine, G. (1996). Angels and devils: Moral landscapes of childhood. *Environment and Planning D: Society and Space, 14*, 581–599.

Vertovec, S., & Cohen, R. (2002). Introduction: Conceiving cosmopolitanism. In S. Vertovec & R. Cohen (Eds.), *Conceiving cosmopolitanism: Theory, context, and practice* (pp. 1–23). Oxford: Oxford University Press.

Werbner, P. (1999). Global pathways. Working class cosmopolitans and the creation of transnational ethnic worlds. *Social Anthropology, 7*(1), 17–35.

Wood, B. E. (2012). Crafted within liminal spaces: Young people's everyday politics. *Political Geography, 31*(6), 337–346.

'FEEL IT': moral cosmopolitans and the politics of the sensed in tourism

João Afonso Baptista

ABSTRACT
Interaction is a matter of concern in all human activities. So far, this basic principle in tourism has been largely analysed and promoted through the perspective of 'the gaze'. In line with a long North-Atlantic tradition that values vision over all the other senses, tourists are too often stereotyped as gazing subjects. In this article, I present tourists in a more encompassing way: as sensing subjects. I contend that the integration of virtues such as morality and cosmopolitanism in tourism derives considerably from the deliberate inclusion of the sensory in tourism activity. These are virtues best authenticated to the tourists through multisensorial incorporation, rather than just through detached gaze. I address the importance of multisensorial experience in the constitution of tourists' cosmopolitan selves in moral terms by drawing on my own ethnographic research in the Mozambican village of Canhane.

It is 08:15 in the morning and I am on an S-Bahn travelling to the neighbourhood of Dammtor in Hamburg. Outside, the sky is dark grey, almost as sombre as the facial expressions and clothing of the crowd that fills the train carriage. The journey passes in silence. Most of the commuters appear unresponsive to the world outside the screens of their smartphones. They seem fed by routine. We stop at Heimfeld, and the couple of passengers standing in front of me leave the carriage. I am suddenly invaded by the vivid colours of the advertisement posted on the dividing wall that had been obscured by the two passengers. The poster depicts a person passionately hugging a *sequoia sempervirens*, a member of the world's tallest tree species, found in coastal California. The advertisement de-homogenises the indoor atmosphere of monotony and gives life to the passenger train carriage. I see various people moving their gaze from their smartphones to the poster. Some of them seem enchanted by it. The enticement of the poster does not come solely from the intense colours and the expressive photo that alludes to a compassionate union between humans and a remarkable nature somewhere out there. Fundamentally, it comes from two words in white capital letters in English, emblazoned across the trunk of the tree being hugged: 'FEEL IT'. Although it is sponsored by a travel agency, the German franchisee Explorer Fernreisen, the poster does not advertise a tourist destination, as it is typical in tourism publicity, but rather personal and direct sensuous involvement in radical difference; the fleshy involvement in worlds beyond everyday sameness which, the hug suggests, are nevertheless worlds that belong to and should be cared for by everyone.

In this essay, I show how the promotion of sensitive engagement with the elsewhere and the different in tourism can be nurtured through a politics of the sensorial. My argument is that the exaltation of immediate sensorial involvement between travellers and the travelled-to or travelled-through by the tourism industry plays a key role in the extension of the role of tourists into that of moral cosmopolitans (I think of moral cosmopolitans as individuals who internalise and imply the global and the plural as part of their moral actions). Yet, in contrast to classical paradigms that promote sensory experience and the morality gained through it as inherently apolitical, undisciplinable, and unmasterable (a domain of instinct), I contend that such experience and morality may be rather the product of deliberate politics of permission and decision. Sensory life is not free of premeditated regulation. More to the point, the morality attained through tourists' sensory experience may be strategically orchestrated by the people and institutions which shape and sanction the actual events of tourist sensation.

Before proceeding, let me clarify the ways in which I use the term 'morality', and the manner in which such a conception of morality is put to use in the service of the cosmopolitan condition in tourism. To be sure, the morality that I approach here is a subjective judgement on the part of the tourists. I am well aware of the dangers of associating human subjectivity – such as morality – with a certain activity and category of actors (i.e. the morality of tourists). Like other anthropologists (e.g. Fassin, 2012; Kean, 2015; Lambek, 2015; Zigon, 2014), I agree that what comes to be felt as morality is contingent upon the particular trajectories of the individual and is never fully stable, and therefore is constantly being reassessed by the encounters one has, the events one participates in, or the unanticipated happenings in one's life. However, as plural and conditional as it might be, the meanings of being moral or acting morally descend from a fundamental ontological starting-point which indicates that morality exists or manifests only because of relations. What I am implying is that all possible types of morality emerge in the context of relations – always. They have significance because we are beings-in-relation: we are always and constantly in relation with something, whether this being human or nonhuman, material or immaterial; and the sense of moral or ethical emerges, receives meaning, and gains purpose by means of such relations. Hence, moralities are not just affixed to the world. Instead, they arise from the relations in it.

In this essay, the morality that I explore pertains to *a* mode of relating to people and things.[1] This is *a* morality constituted through tourists' relationships of bodily immediacy in and with the worlds they attend: in order to flourish morally, tourists must sense directly other ideas, practices, values; they must sense 'in the flesh' and be open to other lives, respectfully, so that they can truly regard themselves as members of a 'world in common'. In practice, the kind of relational morality that I approach here implies that rather than visiting an African village in the comfort of an air-conditioned 4 × 4 car or a sightseeing tour bus, the tourist incorporates morality by walking there, by accepting and ingesting the foods that local residents might share with her/him, and by 'feeling' and participating in local life as if she/he belonged to the village. For these tourists, their moral cosmopolitan condition arises from their propensity to sensorially engage with the other; a propensity that is intrinsic to their identitites and tourism biographies. Hence, the morality that I explore forms through tourists' eagerness to discard their potential hierarchical status and to 'hug' the otherness of diversity in their travels. In a nutshell, I refer to a morality that forms from and expresses through the spirit of being-at-home in and sensory involvement in the plural world.

I organise my arguments through three main sections. First, I introduce the village of Canhane in Mozambique, which has been reported worldwide as an exemplary case of so-called community development in tourism. I briefly familiarise the reader with the specialised tourism that has emerged in Canhane. The processes and dynamics occurring in this village serve as the basis for the general discussion. I demonstrate that tourists attain self-cultivation in Canhane – specifically, they acquire cultural, development, and moral capital – through their unmediated immersion in the locale. Second, I concentrate on the role of sensation in the constitution of morality and cosmopolitanism in tourism. To make my point, I continue with the example of Canhane and describe one of its most successful tourist products: 'the stroll in the village'. This is a walking tour into the village that the tourists pay for in advance at the community lodge where they stay. I aim to demonstrate that what facilitates these touring tourists' recognition of themselves as moral cosmopolitans is not just the money they spend with the intention of helping others 'who have less' but, fundamentally, their orientation towards 'FEEL IT'; that is, their commitment to involve themselves in a diversity characterised by scarcity and material deprivation, directly through their senses. In this view, sensory activity in tourism is a practical way by which individuals apprehend and constitute their moral worlds. Finally, I dedicate the last section to discussing the ways in which the incorporation of moral cosmopolitanism through sensory perception in tourism may be a product of deliberate planning by a specialised industry that combines development with tourism.

Canhane: 'developmentourism' in Mozambique

At the turn of the last millennium, the village of Canhane in Southwest Mozambique was born to the world of tourism – it transformed into a 'destination'. This birth was patronised by the Swiss Nongovernmental Organisation (NGO) Helvetas and the United States Agency for International Development (USAID). After a few years of countless consultations, workshops, and lobbying operations conducted by these institutions, the once-secluded Canhane saw a lodge opening to international tourists on its lands. This lodge, subsequently named Covane Community Lodge, was projected to operate under the values of responsibility and goodness. Among other qualities, it was projected to provide local experiences related to issues of global concern for the tourists, experiences that could then enrich tourists' moral life-worlds. From May 2004 onward, tourists from all over the world could find there a place and a people apparently needing their help and welcoming their participation in the development of the locale.

I conducted anthropological fieldwork in Canhane during 2006 and throughout the whole of 2008. This village, I was told then, has a population of around 1100 residents, who mostly speak Shangane, with only a few being fluent in Portuguese, the national language of Mozambique. The tourist-visitors originate mostly from North-Atlantic countries. Some of them work in Mozambique, holding short-term contracts with NGOs, others came directly from abroad. All of them seem to have anticipatory knowledge of the Covane Lodge and its community-based principles. These are individuals averse to mass tourism, in pursuit of personal sensations in and with the remote unfamiliar.

The major tourism attraction of the village is its expressive material limitations, which the international visitors commonly associate with poverty (Baptista, 2012). But what is most peculiar about Canhane and the Covane Lodge is that they reveal how the industries of development and tourism can melt into each other. Remarkably, development professionals and tourist vacationers are locally perceived as indistinguishable. Both are transient figures that come to the village to help in local development, as well as to monitor the application of their donations and the knowledge they have passed on to the local residents. In Canhane, development and tourism share the same logics of action, the same rationale of intervention, and in practice, are fused into one singular activity. This makes the employment of the conjunction 'and' between the two words unnecessary, if not incorrect. To put it simply, in Canhane there is no development *and* tourism, but only 'developmentourism' (Baptista, 2011).

As soon as the tourists enter the reception office of the Covane Lodge, they encounter several grateful and captivating phrases posted on the walls. Take these two examples: 'Your presence contributes to the improvement of the livelihoods of the population of the village of Canhane' and 'Go and gain more from our village'. These, like many other sentences distributed throughout the lodge, are a sort of 'marketing engagers' – tactical instruments promoting the virtues of engaging with the local(s) – developed and implemented by the NGO Helvetas and its consultants. They suggest that the personal, unmediated experience of local everyday life is what is offered to the tourists. It is by immersing themselves in the local, the catchphrases suggest, by personal involvement in the limitations and potentials of the village that the tourists can contribute 'to the improvement of the livelihoods of the population' and, consequently, can 'gain more' for themselves. This, of course, does not appeal to all kinds of tourists. The tourism order promoted in Canhane is more likely to captivate those who seek 'divergent cultural experiences … for contrasts rather than uniformity', to use Ulf Hannerz's words (1996, p. 103). That is, it is more able to captivate those who can self-realise by 'engag[ing] with the Other' (1996, p. 103) in moral terms than those who look for more familiar or much-travelled environments.

Hannerz's words explain nicely the character of developmentourism in Canhane. Yet he employed them to define a much more comprehensive phenomenon, which he calls 'genuine cosmopolitanism'. Obviously, my adoption of Hannerz's words is not made lightly. With this choice I wish to highlight how Canhane espouses a model of cosmopolitanism. To the point, the international allure of Canhane rests on the possibilities that it offers to international tourists for them to embody and practice the maxim that 'every human being has a global stature as an ultimate unit of moral concern', to use now Pogge's (1992, p. 49) characterisation of moral cosmopolitanism. Accordingly, the Mozambican village emblematises a cosmopolitan framework of goodness in which all human beings, independent of their geographic location, religion, ethnicity, political beliefs, and so forth, must be properly taken into account both in practice and *in loco*. With the emergence of developmentourism in Canhane, this village became a field of simultaneity between socio-economic contrasts and global responsibility. There, tourists can fortify their moral selves by personally engaging in helping the deprived and different local residents as part of a more general process of serving the universal community; a quest which emanates from a sentiment of global belonging. My point is that one essential requirement for such cosmopolitan experience to be morally meaningful, singular, or 'genuine' for the tourists is that they have to 'FEEL IT'; feel in their flesh the unfamiliar and the out-of-the-way lives that they intend to help. In this case, morality and genuineness emerges from the potency of unmediated sensation. How does Canhane accommodate and reproduce such a morality and such a cosmopolitan character? Take the case of a local tour.

The stroll

The Covane Lodge is outside the village. It is located at the shore of the Massingir Dam's lake, around seven kilometres from the area where the population of Canhane lives, and is reachable by a red sand road that passes through an unpopulated forest. There is a prevalent ordered way for tourists to encounter the local residents in their everyday lives, and that is 'the stroll in the village' (originally announced in Portuguese as 'passeio á aldeia').

The stroll is a walking experience that takes tourists to five particular spots in a specific order: (1) the community leader's household; (2) the medicine man; (3) the shallow well; (4) the local primary school; and (5) the water supply tank. This tour is basically a service that can be purchased at the reception of the lodge, and is guided by either the manager or the sub-manager of the lodge. Neither of them have had any training in guiding tourists. Although Portuguese and English are the languages they usually use while touring, Shangane is their first language, which often complicates their ability to communicate. Yet their difficulty with the tour languages reinforces the authenticity of the whole experience for the tourists; it validates the otherness of the locale and the experience in the village as unpretentious and real. Furthermore, the fact that the tour is conducted on foot is crucial since it enables tourists' direct sensation of what they visit. By sensation I mean 'the heterology of impulses that register on our bodies' (Panagia, 2009, p. 2). In this sense, the stroll is more than just an event of perception. It is an experience of *dwelling*. Heidegger (1975) interprets dwelling as a way of being in the world in which one is intimately interwined with and concerned for it. Likewise, the stroll is an occasion for tourists to develop a direct and compassionate relationship with Canhane and its residents, a relationship built from direct embeddedness, and unfolded by their sensing, walking bodies.

The multisensory quality of tourists' walk and their momentary bodily proximity with the materially deprived Other in the village, confirms to the tourists their own venerable value in helping them. Indeed, the affective and humanist force of the stroll does not derive merely from vision or, as has been addressed recently in tourist studies, from the 'embodied gaze' (Edensor, 2006; Obrador-Pons, 2003; Urry & Larsen, 2012). It derives from multisensory and synesthetic practice (Merleau-Ponty, 1962), which exceeds the hedonistic sightseeing so often associated with package tourism. In this spirit, I argue that through the stroll, Canhane becomes a moral site where a transient sense of supranational 'communitas' (Turner, 1969) that encompasses the differences distinguishing the international tourists from the local hosts emerges. To demonstrate my point empirically, let me guide you into the five standard stops made along the stroll.

First stop: the community leader's household

The community leader is the first sight to be 'sensed' (here I mean both to 'make sense' and that 'can be sensed'). This moment represents a symbolic beginning which authenticates the tourists' entrance into the emblematic space of community. Typically, the headman of Canhane greets the tourists with a handshake in which he grasps his right elbow with his left hand while he shakes the hand of the tourist with his right. It is a sign of respect, showing that the arm of the tourist is heavy. This flesh-meeting interactive moment is a wordless way to communicate both courtesy and esteem to the tourists while it promotes an atmosphere of modesty and hospitality. Touch here is used to affirm feelings of fraternisation and to enhance interpersonal involvement. It reciprocally connects and humanises the protagonists of the encounter. After all, as various authors have demonstrated, touching and being touched can promote 'the feeling of being internal' (Martin, 1995, p. 272) to each other. Since the experience of touch implies that one is usually closer to the interacting entity than in cases of hearing, vision, and smell, Ruth Finnegan (2005, p. 19) says that of all the communication channels, touch is the most effective for marking and making relationships. Following the same reasoning, Constance Classen (2005, pp. 1–2) stresses that tactile sensation is the fundamental medium for the expression and production of meaningful acts. In Emmanuel Lévinas's (1967) opinion, touch is actually at the core of the origin (not just expression) of ethical relations. Because it implies direct contact, touch may lead to a more frightful and honest interpretation of the beingness of the Other. More, it bridges the gap of indifference that may separate one person from the other. This is why anthropologists Geissler and Prince (2010, p. 13), for example, conceptualise touch as the primary modality for making ethical relations in Uthero, Western Kenya.

In the tourist imagination, this initial tactile moment in the stroll serves to integrate the tourist into the local society. At some point, the Canhane headman invites the visitors to sit outdoors on plastic chairs, which are usually placed in a circle. These chairs accommodate the tourists, the guide, the community leader, and, eventually, any other men already there. The women from the village sit on the floor. Tourists are encouraged by the momentary silence that follows to direct questions to the headman. Silences, like any other form of communication, must be understood within the borders of a specific context. In this case, 'silence is an absence with a function' (Glenn, 2002, p. 263). It operates as an 'invitational rhetoric' (Foss & Griffin, 1995), where room is provided for the tourists to feel part of the event, rather than simply spectators. 'How many people are in the community?' 'How old are you?' 'What kind of activities do people do here?' 'Why is the name of the community Canhane?' These are some of the questions that emerge from the tacit pressure. At this point, the tour guide translates the tourists' languages into Shangane and vice versa. The headman engages in the conversation with efficiency, often resorting to documents and maps that he keeps at home. He participates in the stroll by representing the traditional authority and the ultimate source of information about the village. Fundamentally, the underlying raw format of the meeting reinforces the sense of meaningful interaction for the tourists: more than just touring or gazing, they find themselves socialising in the village.

Second stop: the medicine man

Next in the stroll, tourists are guided to one of the local medicine men. The village has four medicine men, locally referred to as either *curandeiros* (in Portuguese) or *niangas* (Shangane). There is an informal hierarchy among them based on their age. The oldest is the most requested in his practice, and is also formally consulted in the village whenever there are collective concerns. His importance in Canhane goes beyond the practice of medicine, and he is always the first choice for the stroll.

The encounter with the medicine man is more individualised, less public, and comparatively more intimate than the previous tourist experience. It happens in an indoor atmosphere shared by the tour guide, the tourists, and the medicine man himself. His consulting hut, which accommodates the meeting, is packed with numerous objects related to the practice of medicine. Some of these objects are hanging in the air from hooks attached to the ceiling, others lie on the floor. But the aspect that most immediately strikes the tourists is the near darkness of the place. The insufficiency of luminosity generates momentary feelings of intimacy and occultness. Authors such as Peter Davey (2004), Mikkel Bille and Tim Sørensen (2007) suggest that 'without light, form and space have little meaning' (Davey, 2004, p. 47). Their view descends from the legacies of the En*lighten*ment, and ideologies of civilisation associated with the primacy of vision. As Classen remind us, '[u]sing the visual adjectives "bright" and "brilliant" to mean intelligent only came into vogue during the era of the Enlightenment, when the cultural importance of sight [and light] was on rise' (2005, p. 5). However, as various contemporary ethnographies on blindness show (Hammer, 2013; Kaplan-Myrth, 2000), forms and spaces can gain meaning not only through the eyes (light and vision), but also through other sensory bodily organs. Accordingly, in Canhane, the darkness of the medicine man's room compels a form of meaning-making and generates 'practices of relatedness' (Geissler & Prince, 2010, p. 13) not subjugated to the power of the visual world alone. There, the tourists are mobilised sensorially to consider other ways of treating illness; there, the tourists can display their competence in dealing with and understanding otherness by becoming creatures of total sensation; there, the tourists are revealed to themselves and to others as beings resistant to a modernity in which, as MacCannell (2011, p. 20) presents it, visibility is the organising principle of social life.

Third stop: the shallow well

The shallow well is the first communal place the tourists visit, and it is there and then that they are introduced to the water scarcity problem in Canhane. The visual impact of shortage is strong and authenticates the sense of a poverty. Mostly frequented by local women, the shallow well is probably the most immediate mark that matches the imaginary of 'underdevelopment' in tourists' minds. It collects public everyday life and scarcity into the same place. In this area, there is a hole in the ground. When there is water in the hole, the place acquires a social vitality difficult to feel in other areas. The women go down to the bottom of the hole with their buckets to fetch water. At this moment, there is no individuality to see or presentation to listen to, but only the population of Canhane embedded in a circumstance of shortage to be felt. Usually, tourists and the guide remain silent. 'Verbal sacralisation' (Fine & Speer, 1985) is redundant, while the sensory experience of being there lifts the tourist to an emotional level and legitimates a specific character of knowledge of the place and of the residents. The entire place is naked of trees and shade, which dramatises the precarious conditions of water supply in the village. The aesthetics of dryness makes this the locale in Canhane where water as a basic resource for human life can be most tragedised and viscerally valued by the tourists. The place in itself communicates inhospitableness to the senses.

At the shallow well, the tourists experience and become cognisant of people and situations – the 'hyper-real', as Lorraine Brown (2013) puts it – in a circumstance that is often shocking for them. It is a poverty show that contrasts with visitors' ordinary lives. A 41-year-old British woman stood immobile for around two minutes, seated on a rock under a hot sun, looking at the setting of the place. 'Are you okay?' I asked her. 'When I see these same situations on TV or on the computer screen', she said, 'I'm not close enough, so it's easy to turn off feelings. But now that I'm here … It's impossible to ignore it.' Her tearful comment cannot be disentangled from the actual sensory potency of the moment and place in which it was made. She confirmed what many scholars have suggested: '[t]here has always been a nagging inadequacy around the assertion that one cannot sell poverty, but one can sell paradise. Today, the tourist industry does sell poverty' (Salazar, 2004, p. 92). Canhane verifies it. But Canhane also shows that the evocative power of poverty in tourism is effective mainly through an appeal to the senses. Tourists become organoleptically sensitive to local poverty. That is, in the shallow well, tourists directly experience an appearance of poverty (Baptista, 2012), and this is what touches them deeply.

The first impression that tourists usually verbalise after reaching the locale is 'poor women'. Their perception is not an outcome of residents' strategic enactment. The shallow well is one of the places in the village where women most like to congregate. Among the most obvious reasons for this is that it is a privileged area for them to socialise with each other, without being called lazy by men. Furthermore, it is where they are informed about the latest rumours in the village, speak about their problems, and reinforce links with other women. Hence, their presence at the shallow well during the stroll should not be framed as 'performed authenticity' or 'staged authenticity' carried out for tourism consumption (MacCannell, 1999). This is not to say, however, that the residents do not play a conscious role in this 'poverty moment'. The way they participate in prompting the 'poor women' connotation can be found in their authorisation of the inclusion of the locale in the stroll. In other words, at the shallow well, the residents allow themselves to be sensorily experienced by the tourists in a pre-selected everyday life atmosphere. This does not always happen, insofar as there are places and situations not allowed for tourists (Baptista, 2012). The shallow well is fundamentally a space, a moment, 'an aesthetic of sensation' – as Mark Stranger (1999, p. 270) would call it – sanctioned in the politics of exhibition.

At the shallow well, visitors are suggestively free to participate and be part of the social action; they are implicated in the place as 'dwellers', as relational beings entangled in it (e.g. Watts, 2013). Sometimes this is manifested by them taking pictures, timidly simulating collecting water from the shallow well, and taking the initiative to approach the women close by. Such a corporeality of movement produces intermittent moments of physical proximity, and the structural confines between the residents and the tourists seem to be attenuated through direct interaction. This is a chance for communion between individuals of radically different socio-economic status with poverty as the background.

Hence, at the same time that the place emanates desolation and comfortlessness, it stimulates transient connections between the sensing-tourist and the sensed-Other. It stimulates a momentary transcultural sense of 'communitas' (Turner, 1969), in the sense that it triggers tourists' compassionate bond with what they feel, independently of the level of cultural and economic contrast implied in that process. Soon, the shallow well becomes a geography of affect. 'No matter who they are and what they do, everybody needs water', a French tourist commented softly to herself. 'We all are humans and we all should have decent access to water. Not this. This is not decent'. From the tourists' side what emerges at this moment in the stroll is a general feeling of responsibility and duty toward the Other. Such a sense of responsibility and duty is indifferent to national, regional, or ethnical units, but grounded in the notion of global community. It is part of a morality that advocates 'decency' to 'all of us', in which every person counts as part of a whole of moral consideration.

Therefore, by internalising their predicaments, tourists integrate the deprived Other into a sense of world citizenry and supralocal group membership in which people are responsible for each other. At this stop, if not before, the tourists become more than just transient 'moral tourists' (Butcher, 2003); they are moral cosmopolitans who participate individually and developmentally beyond the nation-state.

Fourth stop: the school

The local school comes next. Here, the children are taught literacy from grades one to seven. According to a text titled 'Impacts in the community of Canhane', which is announced at the reception of the lodge, the village has a 'conventional school room + twenty school desks, and twenty-seven old school desks that were also *rehabilitated with the tourism income*' (emphasis added). At the school, tourists have the opportunity to experience a positive side of the village which – and this is the important point here – is announced as being a direct consequence of their tourism in Canhane.

'Before, children used to attend classes under that big traditional tree', the tour guide said on one occasion while pointing to a *canhoeiro* (marula tree) not far from them. Most of the tourists who I accompanied on the stroll walked to the *canhoeiro* and remained for a few minutes in its shade: they approached the tree to 'FEEL IT'. Their contemplation of the moment and their imagination of what it is to attend classes under that *canhoeiro* are produced through bodily sensory thinking. As happens in all the other locales that the tourists experience in the stroll, this area and moment of presentation is fundamentally an area and moment of sensation.

The context of 'FEEL IT' and the guide's factual discourse highlight the positive replacement of the previous, precarious teaching situation (under the tree) by the new one (classroom). With this, the tourists are introduced to local development, a development that nevertheless is suggested as only being possible because of them – the tourists themselves. This is expressively reinforced when they are led to a sign inscribed at the entrance of the classroom. It says: 'Primary School of Canhane. Enlargement of the classroom. Contribution of the Covane Lodge and of the community. 2005.' The sign associates a concrete materialisation of development with the local tourism project. It explicitly bonds development and tourism through the assertion of the contribution, and thus participation, of tourists in local development. Tourism and the tourists are developmentalised in the same way that development is touristified. As such, the local school reveals an important characteristic of the village: Canhane as a terrain of development-tourism (Baptista, 2011).

Fifth stop: the water supply system

The stroll finishes with the water tank at the north-east corner of the village. The outline of its apparatus contrasts with the majority of the village's landscape. It embodies social betterment. While underlying the value of the water supply for the community, the tour guide often employs special linguistic patterns that sensitise the visitors (e.g. 'We are now looking at another effect of tourism in the community'; 'The community have applied tourism revenues here, and they have built this'; 'The water tank has changed their lives'). Water, the tank, and tourism are rhetorically configured as community development symbols. The imprinting of meaning on the site becomes greatly emotional.

On numerous occasions, tourists expressed a sort of personal relief revealed through their pleased comments. These are some examples I heard: 'Oh, look at this hidden equipment'; 'What a nice surprise'; 'Water for everybody?!'; 'I'm happy now'; 'Beautiful'; 'Well done'. The apprehension that accompanied the walking visitors through most of the stroll has given place to a field of zeal that is rapidly filled. There it is, right there, just in front of them: evidence of development! This is a place of gratification, of optimism's apparition. It is also palpable. Following their impulsive observations, the tourists often approach the water mechanism and 'FEEL IT'. One placed the fingers of her right hand on the wood that supports the tank. Another grasped the tube of aluminium where the taps were installed. Other tourists walked alongside the flexible plastic tube that supplies water from the river, pressing it gently with their hands in different spots. What became evident was that, for these tourists, the importance of that moment, the value of the water mechanism and their own relation/association to it, gained significance (also) through touch.

Tourists' comments and the virtue of the occasion expressed through their tactile contemplation derived less from the single moment when they faced the water equipment, and more from the communal shallow well that they had 'felt' before. Through this sequential journey, they realise a positive evolution of one of the most basic elements for human life: water access. From the shallow well to the water tank, tourists make a sensory-emotional journey from poverty to betterment, from social embarrassment to human dignity, and from a problem to its resolution. Put differently, the arrangement of evidence of improvement towards the end of the walking tour, and the sensorial dimension of actually experiencing such an improvement, provides a sense of climax at the water tank. In this sensorial odyssey, tourists encounter bipolar infrastructural conditions while a particular display about the possibilities of developing the community is promoted. More to the point, in their sequential journey, tourists can experience and confirm their own positive role for the local population.

In contrast to the shallow well, the area around the water tank is often empty of people. The infrastructure stands alone in the scenery with no social vitality around it. The place holds an atmosphere of stillness. Like the other development effect of tourism in the village – the school – the water tank is rarely frequented by adult residents. Tourists question the reason for this: 'Why is no one getting water here?' an Italian man who was accompanied by his girlfriend asked the tour guide. They are told that there have been technical problems with the water mechanism and that it is a temporary situation, soon to be resolved. Although this argument is not uniformly interpreted and passively accepted by all tourists, they usually do not insist on questioning the reasons for its temporary inoperability. Instead, the visitors tend to celebrate the water system as a real and worthy accomplishment of social betterment in the village. Arguably, their sensorial odyssey overcomes the mastery of pure and unemotional 'rationalisation' in their judgements (Beck, 2002, p. 18).

However, behind the physical apparatus there is another version to be told. It is a version that reveals the antithesis of the local improvement. As I explain more comprehensively elsewhere (Baptista, 2010), due to internal ways of ordering the social, the implementation of the water supply infrastructure in that part of the village was doomed to fail even before it was established, and it actually never functioned. Yet, this version is absent from the tourist walking experience; a version that represents neither what the local residents and the ruling development NGO want to show, nor what tourists want to access. This is why the fact the water supply system does not work is irrelevant to the meaning that it represents in the visitors' minds. The aesthetic experience and the 'perspectival sensing'[2] provide the knowledge that tourists want to gain, likewise the knowledge the residents and the developmentourism industry want to provide. In this sense, and extending beyond the last stop in the stroll to the entire walking experience, what tourists directly sense are idealistic versions of the phenomenon that they access. It is through this framed, perspectival, and multisided structure of feeling, in which tourists use their own bodies to see, smell, touch, and hear the residents' insufficiencies as well as the contributions and potentialities for solving those same insufficiencies, that tourists engage in a project of moral cosmopolitanism.

The pursuit of moral cosmopolitanism

The stroll exceeds mere leisure. It provides the tourists with various significances, but most of these are bound up in a logic of interaction and moral participation. The very way of experiencing Canhane in the stroll stimulates it. As Gros observes, 'walking gives you participation' (2014, p. 96): participation in the solidity or softness of the ground stepped upon, in the infrastructures touched along the way, or in the lives of the people met. More than other ways of experiencing places and people, walking foments participation in the immediate profusion surrounding the walker, and it 'helps retrieve the absolute simplicity of presence' (Gros, 2014, p. 67). Therefore, it both communicates an ethical posture and, concomitant with that, allows more effectively the propensity to feel closer, to understand, and to be part of the travelled landscape. 'Walking', Gros concludes, 'lets you *feel it* in an abyss of fusion' (2014, p. 181, 191, emphasis added).

This form of interactional tourism in Canhane, as bodily and immediately sensing the material and human other, informs a broader picture, since what gives it moral imprint is the speciality of doings carried out by the tourists. Ethics in tourism, I contend, is intrinsic to tourist moral action; that is, it is intrinsic to the tourists' actual doings, such as where they go, how they go, what they pay for, who they meet, and how they perform their tourist character while travelling. It is through such actions, through such doings and relationships of commitment to the travelled-to and travelled-through, that the question of what counts as morality and ethics comes to matter for the tourists and tourism. This proposal of the nature of ethics in tourism suggests a shift in the conceptual focus of tourist studies related to ethics: from 'ritual inversion' (Graburn, 1983) to modes of involvement; from touring to dwelling; from written and graphic mediums to the sensory body, and from representations to the non-representational.

Tourists' sense of supranational care and responsibility for the human beings they interact with has as its enabling condition the structuring circumstances reproduced in the places they visit. In the particular case of Canhane, the tourists are prompted to realise their actions as moral agency in consequence of the inclusion of the sensorial in their tourism experience: they enrich their moral selves through direct (I mean *directed*) sensation. For example, the bodily bringing of tourists closer to the new classroom and to the water tank encourages them to value in a visceral, and therefore deeply felt, way the money they spent in the Covane Lodge, as well as their presence and, more utterly, their very beingness. It is mostly through their sensorial activity in the village that the tourists realise themselves as protagonists of a morality that is conspicuously associated with the task of 'helping a community of others'.

Yet in this context, sensing is not only an important facet of lived experience with respect to the other, but also an important moral commodity. Actually, Canhane represents a prime example of the 'experience economy' (Pine & Gilmore, 1998), in which what is sold to the tourist-consumer is inherently personal, individually felt, and specifically memorable. While classic economic products are external to the buyer, experiences are intrinsic to the individual who engages in them on an emotional, bodily, or even spiritual level. This is in line with the emergent rise of sensory marketing, which starts from the precondition that what consumers value most derives from the means of consumption rather than from the products themselves. In tourism, the role of sensory marketing is growing. For instance, the enterprise São Paulo Turismo (SPTuris) in Brazil recently launched the 'map of sensations', which encourages visitors to experience certain pre-identified spots in the metropolis through their senses (mapadassensacoes.com.br). Likewise, the stroll is an attempt to institute a 'sensescape' (to borrow Appadurai's style) in Canhane of moral worthiness through an economy and politics of the sensory in tourism.

However, I do not intend to promote the idea that there are powerful independent entities – the local residents and the developmentourism industry – dominating and inducing another independent and powerless one: the tourists. Developmentourism is not unilateral. Particularly evidenced through the stroll, Canhane is constituted as a meaningful setting that tourists consume, but which they also help to produce. Tourist practice here goes hand in hand with the tourists' quest for cosmopolitan self-cultivation. Indeed, those eager to visit the village seem to share a particular way of experiencing tourism. Their choice to take the stroll is part of their tourist identification with moral sensitivity to the other; an identification mostly built in opposition to the mass-tourist stereotype that, as Hannerz (1990, pp. 241–242) asserts, has no interest in difference. In contrast, the tourists of the stroll present themselves as sensitive to the unfamiliar community's problems, and aspire to feel those problems as part of their personal tourist experience. On one occasion, after visiting Canhane, a Portuguese tourist who stayed two nights at the Covane Lodge told me:

> The tourism agency in Maputo didn't want me to come here. I asked them, 'Why?' They said, 'Because you won't have nice conditions there; it's a poor place, they don't have electricity, hot water, good facilities, and blah, blah, blah ... ' My answer to them was: 'It's precisely because of that I want to go'. I had to prove to them I'm not a typical tourist; otherwise they wouldn't stop talking about that.

A few minutes later, she concluded by saying, 'I want to be closer, *feel*, understand and help the community'. She announced herself in terms of what Martin Zuckerman (2007) calls a 'sensation seeking' personality. Despite the efforts of the tourism agency, she stuck by and defended her choice, and in so doing she manifested and reproduced her idealised tourist status. In her case, it is clear that Canhane and the stroll have significance as moral destination and practice, which in turn serves to sustain or cultivate a self open to plurality and which competes for distinction from the ordinary version of tourists. At stake, here is the sense of involvement through tourism in a sort of lifestyle that goes in opposition to monocultural frameworks; a rejection of homogeneous, insensitive, careless ways of being-in-the-world.

While embracing a benign tourist *persona*, this Portuguese traveller placed herself as a person who could 'help them', and therefore as someone who perceives herself to be different from the local residents. In this case, one could say that developmentourism in Canhane is a 'business of "difference"' (Salazar, 2004, p. 85), which means it is an arrangement that favours those who seek to incorporate the cosmopolitan condition. Embracing difference, openness to other ways of life, and integration with the unfamiliar are at the core of tourism in Canhane. This became evident to me when listening to the often lengthy 'tourist biographies' told by the tourists themselves at the Covane Lodge. As one German visitor said,

> I often travel to places like this [Canhane]. Among other reasons, I love community tourism in Africa and South America because it implies entering a world of differences. I see and *feel* different things and different ways of living. I love it.

Like many other tourists visiting the village, the appeal of contrasts and the impulse to feel such contrasts directly motivate this tourist. He explicitly seeks to immerse himself in a positive conception of a pluralistic, bounded world: 'just love it'.

What this means is that when tourists visit Canhane they are not passive elements acting naively as human puppets, or gullible consumers of set-up sensorial dispositions: they also seek to 'FEEL IT' and to acquire moral worth, to that end resorting to a sort of leisure catharsis. In this sense, the stroll can be characterised by tourists' preceding desires to feel realities different from their own – including poverty – and by the active role of the residents in providing it. The stroll reflects and confirms the tourists' aspirations and idealised self-distinctive lifestyles of moral cosmopolitanism back to them. Thereof, it incorporates a strong component of ego-touring (Munt, 1994): it is a way for tourists to accumulate a particular type of cultural and moral capital that supports their idealised identifications, their expectations, and their worldviews. It is a site of tourists' personal, cosmopolitan, and moral development.

Often the tourists of the stroll leave the site somehow involved in a momentary conscience of doing good and, as one tourist told me after she finished the tour, they 'take something meaningful' with them. Naturally, the enjoyment and gratification that these tourists obtain implies that they are necessarily open to or interested in material shortage and personal integration in diversity. In the end, the stroll and the entire developmentourism activity is interwoven with tourists' significances and investments. The moral cosmopolitan modes of being of the tourists are at the very heart of this principled modality of touring and tourism.

The political in sentient existence

In Canhane, the tourists are the ones who opt to be guided into the village rather than visit it by themselves. Through their preference, they put their sensing interactions at the mercy of being managed. To say this is to contend that 'the sensoriality of claim making' (Panagia, 2009, p. 17), which derives its principle of authenticity and ethics from the *being in* the phenomenon of appreciation, can be subjected to a preconstituted referential model. Sensory life is indeed susceptible to being guided. This is particularly evident in the stroll, when certain sensory practices that fortify tourists' quest for self-cultivation, self-worth, and moral cosmopolitanism are exalted, while the other practices that could frustrate such a quest are not enabled. Here, I am particularly influenced by Jacques Rancière's concept of the 'partition of the sensible':

> We will call 'partition of the sensible' a general law that defines the forms of part-taking by first defining the modes of perception in which they are inscribed ... This partition should be understood in the double sense of the word: on the one hand, that which separates and excludes; on the other, that which allows participation. A partition of the sensible refers to the manner in which a relation between a shared 'common' and the distribution of exclusive parts is determined through the sensible. This latter form of distribution, in turn, itself presupposes a partition between what is visible and what is not, of what can be heard from the inaudible. (2001)

Rancière explicitly places sensory experience under the aegis of a politics of separation and exclusion, which in turn can shape our understandings of the human and nonhuman other. If we see the sensorium as constitutive of subjectivity, as Tønder (2015) stresses, then this process of 'partitioned' provision and accessing of sensation – what might be called a politics of selection of what is to be sensed – gains crucial relevance. This is because, to a great extent, the ideas and meanings that have effects in individuals and societies originate from the selection of what can or cannot be sensed. Precisely, in Canhane, it is the stimulation of tourists' multi-sensory capacities in *presanctioned* encounters of sensation that helps to constitute a series of key subjectivities for them, such as their sense of moral participation in local development, self-distinction from mass-tourists, and their very notion of difference upon which they build their identities and exercise their moralities of cosmopolitanism. The validation and authentication of Canhane as a society to be morally developed via tourism is greatly reliant upon the efficacy of a politics over the tourists' senses.

Conclusion

In this essay, I present the walking tour in the Mozambican village of Canhane as a tourist experience of *dwelling* (Heidegger, 1975) structured in economic asymmetries, which is supersaturated with significances sensorially apprehended. This event of sensation is a planned ceremony of intention enabled by the organisation of the sensed. It occurs under a prearranged disposition that orders what comes to count as subjects of perception.

Hence, my main goal is to address how central the particular ways of inclusion of the sensory in tourist experience can be to the constitution of tourists' cosmopolitan selves in moral terms; or, better put, to show that the character of benevolence and cosmopolitanism in tourism can be related to specific ways of experiencing destination landscapes and societies, namely through walking, smelling, listening, touching, and directly sensing. In contrast to common assumptions that equate cosmopolitanism with metropolitan residence, travel, commodity consumption, and the instantaneity of accessing messages from all over the globe (i.e. in the smartphones of train commuters in Hamburg), Canhane, its residents, the tourists who visit it, and the developmentourism industry demonstrate that the self-awareness of belonging and contributing to a more 'genuine' (Hannerz, 1996) and moral form of cosmopolitanism occurs primarily out of direct sensation: 'FEEL IT'.

Unmediated sentient experience, however, is subjected to what Davide Panagia (2009) calls 'a regime of perception'. In Canhane, with the distribution and assignment, inclusion and exclusion of the sensed, the stroll structures the tourists' modes of attending to the village. This distribution organises not only what is to be sensed but also what is not; what can and cannot be heard, smelled, and be seen (Baptista, 2012). In this vein, the stroll organises the ways and possibilities for the tourists to constitute subjectivities concerning morality, identity, selfhood, justice, difference, poverty, and development. To put it differently, Canhane shows how the promotion of 'FEEL IT' as a vehicle for the individual to access or get closer to the cosmopolitical and moral status does not come without broader power implications.

The circumstances of Canhane and the stroll suggest that more attention should be paid to the corporeal and non-representational engagements that actually happen in tourist environments. Being as it is a privileging field to access the ideological constitutions of ourselves and the Other, I argue that exploring the politics of 'FEEL IT' in action is an important step towards a better understanding of how we all (whether as 'tourists', 'citizens', 'cosmopolitans', 'locals', 'scholars', or otherwise) conceive of important subjects, such as morality, difference,[3] ourselves, and others. Specifically in tourism, and in contrast to mainstream understandings of tourists as gazing subjects (Urry, 2002), I therefore suggest interpreting tourists firstly and fundamentally as *sensing subjects*.

Notes

1. The emphasis on the indefinite article 'a' serves to reply to one reviewer who noted that one does not need

immediate sensatory experience to engage in moral acts. As the reviewer wrote, 'I can approach morality via reading'. Indeed, I do not argue against this claim. But in this article, I focus on *a* different type of relationship with things and people, *a* relationship not so much mediated by text books, photos, or narratives, but characterised by corporeal proximity and immediacy.
2. By mentioning 'perspectval sensing', I expand on Paolo Favero's (2007, p. 57) concept of 'perspectival seeing', and thus extend the perspective of tourist selectivity beyond vision.
3. It is now well known that since the Enlightenment, 'difference became the primary organizing principle' of perception, action, and culture (Langerman, 2012, p. 177).

Disclosure statement

No potential conflict of interest was reported by the author.

References

Baptista, J. (2010). Disturbing 'Development': The water supply conflict in Canhane, Mozambique. *Journal of Southern African Studies, 36*(1), 169–188.
Baptista, J. (2011). The tourists of developmentourism – representations 'From Below'. *Current Issues in Tourism, 14*(7), 651–667.
Baptista, J. (2012). Tourism of poverty: The value of being poor in the nongovernmental order. In F. Frenzel, K. Koens, & M. Steinbrink (Eds.), *Slum tourism: Poverty, power and ethics* (pp. 125–143). London: Routledge.
Beck, U. (2002). The cosmopolitan society and its enemies. *Theory, Culture & Society, 19*(1–2), 17–44.
Bille, M., & Sørensen, T. (2007). An anthropology of luminosity: The agency of light. *Journal of Material Culture, 12*(3), 263–284.
Brown, L. (2013). Tourism: A catalyst for existential authenticity. *Annals of Tourism Research, 40*(1), 176–190.
Butcher, J. (2003). *The moralization of tourism – sun, sand … and saving the world?* London: Routledge.
Classen, C. (2005). Fingerprints: Writing about touch. In C. Classen (Ed.), *The book of touch* (pp. 1–9). Oxford: Berg.
Davey, P. (2004). Light and dark. *The Architectural Review, 215*(1286), 46–47.
Edensor, T. (2006). Sensing tourism spaces. In C. Minca & T. Oakes (Eds.), *Travels in paradox: Remapping tourism* (pp. 23–46). Lanham: Rowman & Littlefield.
Fassin, D. (Ed.). (2012). *A companion to moral anthropology.* Sussex: Wiley-Blackell.
Favero, P. (2007). 'What a Wonderful World!': On the 'Touristic Ways of Seeing', the knowledge and the politics of the 'Culture Industries of Otherness'. *Tourist Studies, 7*(1), 51–81.
Finnegan, R. (2005). Tactile communication. In C. Classen (Ed.), *The book of touch* (pp. 18–25). Oxford: Berg.
Fine, E., & Speer, J. (1985). Tour guide performances as sight sacralization. *Annals of Tourism Research, 12*(1), 73–95.
Foss, S., & Griffin, C. (1995). Beyond persuasion: A proposal for an invitational rhetoric. *Communication Monographs, 62*, 2–18.
Geissler, P., & Prince, R. (2010). *The land is dying: Contingency, creativity and conficlt in western Kenya.* Oxford: Berghahn.
Glenn, C. (2002). Silence: A rhetorical art for resisting discipline(s). *JAC, 22*(2), 261–291.
Graburn, N. (1983). The anthropology of tourism. *Annals of Tourism Research, 10*(1), 9–33.
Gros, F. (2014). *A philosophy of walking.* London: Verso.
Hammer, G. (2013). 'This is the anthropologist, and she is sighted': Ethnographic research with blind women. *Disability Studies Quarterly, 33*(2). Available at http://dsq-sds.org/article/view/3707/3230
Hannerz, U. (1990). Cosmopolitans and locals in world culture. *Theory, Culture & Society, 7*(2), 237–251.
Hannerz, U. (1996). *Transnational connections: Culture, people, places.* London: Routledge.
Heidegger, M. (1975). *Building dwelling thinking.* New York, NY: Harper Colophon.
Kaplan-Myrth, N. (2000). Alice without a looking glass: Blind people and body image. *Anthropology & Medicine, 7*(3), 277–299.
Kean, W. (2015). *Ethical life: Its natural and social histories.* Princeton, NJ: University of Princeton Press.
Lambek, M. (2015). *The ethical condition: Essays on action, person, and value.* Chicago, IL: University of Chicago Press.
Langerman, F. (2012). Trees, webs and explosions: The analogical imperative in the politics ok knowledge. In S. Levine (Ed.), *Medicine and the politics of knowledge* (pp. 173–186). Cape Town: HSRC Press.
Lévinas, E. (1967). *En découvrant l'existence avec Husserl et Heidegger.* Paris: Vrin.
Martin, M. (1995). Bodily awareness: A sense of ownership. In J. L. Bermúdez, A. Marcel, & N. Eilan (Eds.), *The body and the self* (pp. 267–289). Cambridge, MA: MIT Press.
MacCannell, D. (1999). *The tourist: A new theory of the leisure class.* Berkeley, CA: University of California Press.
MacCannell, D. (2011). *The ethics of sightseeing.* Berkeley: University of California Press.
Merleau-Ponty, M. (1962). *Phenomenology of perception.* London: Routledge and Kegan Paul.
Munt, I. (1994). Eco-tourism or ego-tourism? *Race Class, 36*(1), 49–60.
Obrador-Pons, P. (2003). Being-on-Holiday: Tourist dwelling, bodies and place. *Tourist Studies, 3*(1), 47–66.
Panagia, D. (2009). *The political life of sensation.* Durham, NC: Duke University Press.
Pine, J., & Gilmore, J. (1998). Welcome to the experience economy. *Harvard Business Review, 76*(4), 97–105.
Pogge, T. (1992). Cosmopolitanism and sovereignty. *Ethics, 103*(1), 48–75.
Salazar, N. (2004). Developmental tourists vs. development tourism: A case study. In A. Raj (Ed.), *Tourist behaviour – A psychological perspective* (pp. 85–107). Delhi: Kanishka.
Stranger, M. (1999). The aesthetics of risk: A study of surfing. *International Journal for the Sociology of Sport, 34*(3), 265–276.

Tønder, L. (2015). Political theory and the sensorium. *Political Theory*. doi:10.1177/0090591715591904

Turner, V. (1969). *The ritual process: Structure and anti-structure.* Chicago, IL: Aldine Publishing.

Urry, J. (2002). *The tourist gaze*. London: Sage.

Urry, J., & Larsen, J. (2012). *The tourist gaze 3.0*. London: Sage.

Watts, C. (Ed.). (2013). *Relational archaeologies: Humans, animals, things*. Abingdon: Routledge.

Zigon, J. (2014). An ethics of dwelling and a politics of world-building: A critical response to ordinary ethics. *Journal of the Royal Anthropological Institute, 20*(4), 746–764.

Zuckerman, M. (2007). *Sensation seeking and risk behavior.* Washington, DC: American Psychological Association.

Cosmopolitan education, travel and mobilities to Washington, DC

Felix Schubert and Kevin Hannam

ABSTRACT

This paper examines the cosmopolitan mobilities of young elites that take part in study-internship programmes in Washington, DC, US. In the case of Washington, DC, a large study-internship industry has been developed and this is an important example of how cities can become instrumental in organising specialised elite mobilities. These study-internship programmes (normally called Washington Semester Programmes (WSP)) give both US and international students the chance to study and intern in Washington, DC. Similar programmes exist in many global cities; however, Washington, DC has arguably become a central hub for those who wish to pursue careers in the fields of development politics or in the NGO sphere. The paper illustrates how ideas and stories of mobile careers and the importance of 'being mobile' on the job market catalyse student mobility into Washington, DC. Significantly, student mobilities to Washington, DC combine education with aspects of tourism and lifestyle mobilities. Moreover, these programmes allude to ideas of global citizenship through increasing participant's human capital by enhancing their cosmopolitanism through this educational experience. Likewise, the participants in these programmes buy into those ideas of cosmopolitanism and the added value to their mobility capital through experiencing the political landscapes of Washington, DC.

Introduction

With the election of President Donald Trump, Washington, DC has become the focus of current media attention. Washington, DC is well established as one of the most important centres of power in the Western Hemisphere and mostly owes this reputation to its role as the capital of the US as well as being the residency of US presidents. For many tourists that is about the extent of the city's image. According to the US Census Bureau, the Washington, DC metropolitan area has an estimated 6 million inhabitants and a student population of over 450,000 (Erickson, 2012). It is the most educated and regarded as the most affluent metropolitan area in the US (Marchio & Berube, 2015). According to Trujilo and Parilla (2016), 48% of the population had tertiary education degrees. In 2014, tourism to Washington, DC set an all-time record with over 20 million visitors, partly due to a 16% increase in international visitors over the previous year (Reuters, 2015). Around 90% of the city's visitors, however, still come from within the US (Erickson, 2012).

Since the 1990s, Washington, DC has experienced ongoing gentrification and ethnic and racial transformation (Jackson, 2015; Knox, 1991; Maher, 2015). Moreover, for many US residents, Washington, DC and everything within the beltway (physically embodied by the Interstate 495 that encircles Washington, including parts of Maryland and Virginia) stands for an elitist sphere of influence:

> 'Inside the Beltway' is an expression we Americans hear all the time, yet routinely I'm asked what it means. Geographically, it's everything within the capital beltway, a sixty-six-mile loop of deadly asphalt that, when not at its customary standstill, carries speeding motorists around Washington. But more often it refers to a mindset, or a malady. A person inside the Beltway can be devoid of common sense, on the take, out of touch with reality—out of touch with America. (McCaslin, 2004, p. 77)

It is significant that, in this comment, the beltway symbolises a spatial limitation and also a mindset that the author describes as being perhaps out of touch with the everyday reality of many US residents as well as visitors including students and tourists. In a subsequent interview, McCaslin described how you 'get caught up in Washington and all the politics, all the shenanigans, and it's like a syndrome' (C-SPAN, 2004).

In the American election of 2016, Donald Trump was able to gain the support of many American citizens by a rhetoric that included many attacks against the elites and political establishment. As populism is on the rise, there have been growing resentments against elites and especially with regard to Washington and the US.

There are the terms 'beltway politics' and 'inside the beltway' which both stand for the ruling elites in Washington, DC. Nonetheless, 'getting into Washington', being able to live in Washington, and having a successful career in Washington seems to be something that has been, and still is, attractive for many young people from all over the world.

Standish argues that the growing demand for global education and education in global citizenship is rooted in 'a sense that the world has changed: that we are no longer living in homogenous communities bounded by national borders, but rather than we inhabit a global society that has placed new demands upon individuals' (2012, p. 1). Moreover, he explains that the following values are increasingly taught and demanded:

> New global realities are highlighted including the global market place, multicultural communities, and postnational politics. This global world is characterized by change and uncertainty, brought about by social and economic forces beyond the control of the nation state, and so knowledge and skills very quickly become outdated. In this fluid environment, knowledge of the past, the subject-based academic curriculum, is presented as less important than the skills for acquiring knowledge and working with others. Therefore, we are told, students need a different kind of education. (Standish, 2012, p. 2)

The Washington Semester Programmes (WSPs) discussed below are an exemplar of such specialised educational study-internship programmes that emphasises these skills and values.

This paper thus examines the cosmopolitan mobilities of young elites that take part in study-internship programmes in Washington, DC. Unlike North-South Volunteering literature, there is little literature on study-internship mobilities to and within the global North and this paper aims to analyse these privileged mobilities. We argue that students come to Washington to take part in these WSPs not for the subject-based education but in order to develop elite contacts and showcase their global citizenship. In the case of Washington, DC, a significant study-internship industry has been developed and this is an important example of how cities can become instrumental in organising specialised elite educational and tourism mobilities.

Educational mobilities, tourism and cosmopolitanism

From a mobilities perspective, tourism is seen as integral to wider processes of economic and political development processes and even constitutive of everyday life (Hannam & Knox, 2010). It is not just that tourism is a form of mobility like other forms of mobility such as commuting or migration but that different mobilities inform and are informed by tourism (Sheller & Urry, 2004). Thus, we need to continually examine the multiple mobilities in any situation: mobilities involve the movement of people such as students as tourists, but also the movement of a whole range of material things as well as the movement of thoughts and ideas – including educational ones (Allen-Robertson & Beer, 2010; Hannam & Guereno-Omil, 2015; Williams, 2006).

The mobilities paradigm also calls for a shift of focus, a more in-depth look at the process of mobility itself and the circumstances in which mobilities takes place, maybe constituting the most innovative component of the mobilities paradigm (Adey, 2010, pp. 36–37). As Adey, Bissell, Hannam, Merriman, and Sheller (2013, p. 21) state, 'Mobilities, cultures and identities can best be approached through an attention to routes and paths, flows, and connections'. An essential idea to understanding the purpose of the mobilities paradigm is that mobility has to be interpreted in more than 'its usual connotation – movement' (Adey, 2010, p. 34). Because movements always take place within a framework and have multiple consequences, to reduce their meaning to the sole act of a move from A to B is not adequate. Oftentimes, mobility is just stripped of its meaning by interpreting it purely as the study of movements, therefore making it a more descriptive field of studies: thus, 'mobility is movement imbued with meaning' (Adey, 2010, p. 34).

A great deal of mobilities research has analysed forms and experiences of embodied travel involving the blurring of spaces of work, leisure, family life, migration and, indeed, education, organised in terms of contrasting time–space modalities (ranging from daily commuting to attend university or a once-in-a-lifetime round the world trip) (Hannam, Sheller, & Urry, 2006). In particular, the concept of lifestyle mobility has been developed to describe 'the spatial mobility of relatively affluent individuals of all ages, moving either part-time or full-time to places that are meaningful because, for various reasons, they offer the potential of a better quality of life' (Benson, O'Reilly, & Kershen, 2009, p. 2). As Benson et al. (2009, p. 5) emphasise, the belief 'that spatial mobility in itself enables some form of self-realization' is key to understanding the concept of lifestyle mobility. The concept of lifestyle mobilities has been used to describe the mobilities of people that want to escape everyday lifestyles dominated by consumerism and materialism (Benson et al., 2009, p. 4). The desire for alternative lifestyles in the context of elite transnationalism and self-development has also been widely discussed in various contexts (see, for example, Benson et al., 2009; Cohen,

2010; Cohen, Duncan, & Thulemark, 2015; Rickly, 2014; Thorpe, 2012).

For many students, internship and related volunteering experiences have become ever more popular and a means to raise their social and cultural capital. It has been suggested that volunteer tourism mobilities 'represent a novel (or at least evolving) form of globalised work practice that is bound into the changing needs of the global economy in general – and transnational firms in particular' (Jones, 2011, p. 532). Intercultural experiences (often from the Global South) and internationally recognised qualifications (always from the Global North) are increasingly being expected by transnational firms, and non-state actors such as NGOs, educational providers and volunteers seem to clearly understand this and 'are increasingly aware of and motivated by the specific and hard-to-acquire values, knowledges, skills and attitudes that international voluntary work experience provides' (Jones, 2011, p. 532). Jones (2011, p. 532) further suggests that employers seeking 'these skills and capacities were seen as intangible and different to those young volunteers would acquire from formal education, as well as being only acquirable by working abroad (i.e. outside of their home country)'.

The extent to which the acquisition of these intangible skills also plays a role in internship mobilities has yet to be analysed however. While there have been critical analyses of volunteer tourism (see, for example, Butcher & Smith, 2015; Mostafanezhad, 2013, 2014), research on study abroad and internship programmes have been mainly focused on the educational and cultural benefits of such programmes with little recourse to the wider political ramifications (see, for example, Lam & Ching, 2007; Root & Ngampornchai, 2013). Hence, in this paper, we want to discuss how ideas of global citizenship, as well as stories and perceptions of Washington, DC, mobilise students to the city as young cosmopolitan consumers. Furthermore, we also note how their mobilities have helped to change both the image and fabric of the city.

Research methodology

This paper is based upon qualitative data collection which took place in Washington, DC during 2015. This included interviews with various stakeholders (5), students (19), field observations of urban change in Washington, DC in 2015, as well as analysis of websites and marketing materials from various stakeholders. The stakeholders interviewed were either WSP coordinators, staff or higher officials within the programmes and they were interviewed about their respective programme, Washington, DC and views on Higher Education policies. The four main themes in the interviews with WSP students focused on their lives in Washington, DC, their participation in their respective programme (as well as their decision-making process for participation in the programme), their experiences on the job market as well as their travel biographies. All of the interviews were transcribed verbatim and inputted into the software NviVo for analysis. The websites we analysed used a purposive sampling method using the most prominent and most established providers of study-internship programmes. While we did do a comprehensive review of the websites of educational providers in Washington, DC, we chose to use those that attract the most students.

We then used textual analysis (Hannam & Knox, 2005). Hannam and Knox explain that textual analysis is a 'qualitative technique concerned with unpacking the cultural meanings inherent in the material in question' while the researcher has to draw upon his or her 'own knowledge and beliefs as well as the symbolic meaning systems that they share with others' (2005, p. 24). This analytical method requires the researcher to deal with the collected data and the text very closely, and even more importantly, it requires the reflexivity of the researcher in order to maintain the validity and credibility of the research. This means that the researcher needs to keep assumptions and preconceptions in check and highlight their impact on his or her research, as well as carefully explaining the steps that were taken in the data analysis (Hannam & Knox, 2005).

The Washington, DC internship industry

In Washington, DC, a number of study-internship programmes were developed in the course of the last century. These programmes usually combine internships with subject-specific study courses (at American University, you can take part in programmes on American Politics, Global Economics and Business, Foreign Policy, International Environment and Development, International Law and Organizations, Journalism, Justice and Law, The Middle East and World Affairs, Peace and Conflict Resolution and Transforming Communities). There are a variety of educational institutions that offer study-internship programmes in Washington, DC, including the universities based in DC that offer these programmes, branch campuses of universities that are located somewhere else in the US as well as private non-university organisations.

The largest universities in the Washington, DC area that have significant study-internship programmes are American University (with about 700 participants per year and more than 40,000 alumni since it was founded in 1947 (American University, 2014, p. 2), Georgetown University (founded in 1789) and George Washington University (founded in 1821). These programmes are open to both American and international students (who can fulfil the admission requirements). Then there are also off-branch campuses of universities that are not based in Washington (e.g. the Universities of California System). In 1982, the University of California, Irvine set up the UCDC programme that claims to have more than 10,000 alumni; other examples are Harvard and Stanford who have set up their own programmes in Washington. Stanford University bought a property in north western Washington in 1988 and consequently set up their own programme with about 1300 alumni to date (Stanford in Washington, 2016; UCDC, 2016). There is no official register for these programmes, which makes it difficult to differentiate between universities that physically built off-branch campuses and those that just cooperate and affiliate with existing programmes. Nonetheless, the fact that a significant number of universities offer their students the chance to participate in study-internship programmes in Washington, DC speaks for the success and the demand for this kind of student mobility.

There are also some non-university actors such as The Fund for American Studies, which was established in 1967 and claims to be 'a leader in educating young people from around the world in the fundamental principles of American democracy and our free market system' (DC Internships, 2016). There are also other funds, associations or organisations such as the Washington Center (founded in 1970, with '140 professional staff, associate faculty and Alumni in Residence, 1,600 interns plus several hundred seminar participants each year' and about 50,000 alumni (Washington Center, 2016), and the Washington Internship Institute (established in 1990, 2500 alumni). In addition to all of these programmes, there are summer schools and internship placement programmes organised by universities from outside the US that operate with similar aims.

There are broad estimates that in total about 20,000 interns come to Washington each summer, of whom 6000 intern in Congress (Politico, 2009). According to Johnson (2010), the annual number of interns in Washington ranges from 20,000 to 40,000, of which about 2500 interns are participants of placement programmes. Johnson (2010) estimates that over the past 40 years, 'the programs have collectively placed more than 60,000 interns. Some of them participate in alumni networks that function like college alumni associations, fundraise for the programs, join Facebook groups, volunteer to mentor or take on interns of their own'.

As its website states, American University's WSP (in Washington, DC) is described as an 'academic experiential learning programme', established in 1947, enabling students to 'spend a semester or an academic year in the dynamic, cosmopolitan city of Washington, DC, where you will have access to some of the most influential people and organizations in the world' (American University, 2016). Furthermore, at their internships, which are a part of the programme, students are told that they will 'gain invaluable work experience through an internship at a local organization and meet the movers and shakers of Washington, D.C.' (American University, 2016). While there are increasingly more programmes (both in Washington, DC and in other global cities and hubs of education), the WSP is one of the older programmes and is deeply embedded into Washington, DC's political landscape and was hence chosen as the main focus of this research. In the section below, we discuss the promotion of the study/internship programmes in Washington, DC.

Promoting cosmopolitanism and global citizenship in Washington, DC

Apart from the unique study and networking opportunities of participation in one of these programmes, increasing one's cosmopolitanism and global citizenship as part of the global knowledge economy plays a significant part in the promotion of these programmes. As both, cosmopolitanism and global citizenship develop in global cities, the global impact of Washington, DC is strongly advertised and moreover, the cultural and touristic opportunities of the city are highlighted. For instance, the Osgood Centre, a not-for-profit educational foundation, describes Washington, DC as an intern city where youth and power meet (Osgood Centre, 2016):

> If there is an internship capital, it is Washington DC. If there is a city where youth have extraordinary power, authority, and influence, it is Washington DC. The District of Columbia is host to thousands of interns each semester and tens of thousands in the summer. It is an extraordinary place to network, to make new friends, have once-in-a-lifetime experiences, and to watch (or be a part of) history in the making. With one of the best educated populations in the world, Washington is a place where you begin to synthesize all you learned from your college education and recognize the alternative paths to your future leadership endeavo[u]rs.

The opportunity to intern and live in Washington is clearly marketed as a once-in-a-lifetime opportunity to grow both as a person and career-wise. The way that Washington, DC is described as one of the best-educated populations in the world suggests that it is in fact more than a city but a space that holds the qualities of future leadership and ambition. One could interpret this space as a key node in globalisation that breeds and furthers cosmopolitan capital. The sentence that refers to 'thousands of interns each semester' has a variety of functions. It makes the reader aware of his or her competition but simultaneously raises awareness for this 'special' opportunity to watch or 'be part of history in the making'. Moreover, it soothes young students who might be scared and intimidated by this rhetoric of power and influence, arguing that they are following in the footsteps of others who have started as interns in Washington. After all, they are coming to the 'internship capital'. Clearly, cosmopolitanism and global citizenship are values that are reflected and utilised in this quotation.

American University's advertisement materials for their WSP also emphasise Washington's cosmopolitanism, *pace* and influence. The programme states that Washington, DC is:

> more than the dynamic and cosmopolitan city that is home to President Obama and your U.S. Senators and representatives. It's an international cultural cent[re] loaded with opportunity and teeming with go-getters anxious to share life experiences, debate the day's most timely topics, and weigh in on policies that help shape the world we know. The DC population is savvy and the pace is faster here, but if you can jump in and hang on there's no better place to discover what you're made of. (American University, 2010, p. 2)

There is a certain tone of warning in this quotation, as it alerts that the DC population 'is savvy and the pace is faster here' but this test will show participants of the programme whether or not they are ready and prepared for such an environment. In this cosmopolitan atmosphere of the city, opportunities may come for those who work hard and are ready for this city and who are prepared for a fast pace of life and thus be highly mobile. This quote can actually be read as an updated American/Global Dream of hypermobility and visualises the image of a moving train, pulling away from the observing student who is seeking opportunities.

While American University's WSP hints at the cultural opportunities of Washington and focuses more on the career aspects of participation in the programme, the Washington Center promotes Washington, DC as a touristic city:

> At The Washington Center, you get not only great work and learning experience but also great life experience. Living in the U.S. capital is like nothing else in the world. The city's energy is remarkable at both work and play. There's so much to see and do, and it's all at your doorstep as a TWC intern. (Washington Center, 2015, p. 12)

Here, the exclusivity of the chance of being able to live in Washington is emphasised and it is asserted that it can compare to nothing else worldwide. Thus, the opportunities of Washington, not only for one's career but also personally, as life experience are marketed. They elaborate more specifically that:

> Washington offers impressive architecture and monuments, incredible museums, World-class theatre, great nightlife, a rich international community and restaurants with a wide range of cuisines. Throughout your time with TWC, you'll experience the city in a way that tourists never could. Best of all, you'll get to know fellow students from the United States and around the world. You'll participate in a variety of social activities, trips and adventures together. And by the time the program concludes, you'll have created friendships that remain strong for many years in the future. (Washington Center, 2015, p. 12)

The contradictory aspect of becoming 'more than a tourist' (whilst ostensibly doing tourism nevertheless) that can be found here in this quote are essential to branding the participation in this programme. The study programmes argue that participants will have more of an experience, a better, more sustainable and worthwhile experience than as a tourist, because participants are there for a longer amount of time and are able to utilise recommendations from locals and programme staff (an aspect of authenticity that companies such as Airbnb have promoted and utilised commercially). The networks that are formed in these 'adventures' in Washington, DC are then seen to lay a foundation for further travel and networking, as it is argued that the friendships and networks created may well be international and longstanding rather than just 'weak ties', according to the Washington Center advertisement. The possibility of adventure and developing social networks is, of course, commonly emphasised in volunteer tourism study-internships to the Global South (Mostafanezhad, 2013). A key difference from such volunteering opportunities is the political rhetoric espoused by the education providers in Washington, DC.

The examples above from some of the largest educational providers in Washington, DC emphasise the allegedly rare opportunities the students may acquire in participating in these programmes and thus getting 'into' the Washington, DC political network. The rhetoric works in order to cast DC as a space of political globalisation, hence something common for today's students, but also something fleeting, something that moves and possibly overtakes them and a chance that they will not get a hold of. The space of Washington, DC is described as the other, an extreme out of the ordinary as its benefits and its connections to the world (as a key node in globalisation) and to the decision-makers and elites who inhabit this space are highlighted. The language used emphasises the uniqueness of the opportunity to get into this space of global decision-making and thus furthers the reification of global ideology at the expense of social resistance (Chandler, 2009). Below, we discuss the student's reflections on their movements to and within Washington, DC firstly in terms of their need to build cultural capital in the context of an increasingly competitive labour market, and, secondly, in terms of their mobilities that are amplified through their engagement with their transient experiences of the city of Washington, DC.

Student mobilities to Washington, DC

Building competitive cultural capital

One of the interviewees was a 22-year-old intern for a Congresswoman at the time of the interview and he was extremely concerned about his professional future. Moreover, he stated that he did not enjoy the internship that much because of a lack of responsibility but that this was not that important because 'Here's what is great about it though, even though a lot of what I am doing, I am not enjoying, it still looks good in a resume; as much as I hate to admit that it is the truth' (Interview with Martyn, 2015). He also asserted that he was scared of the job market due to its competitiveness:

> Terrified, it is super competitive. I mean, yeah you have kids going to Stanford, you have kids going to all the Ivy League schools, you know there are so many great schools out there and so many smart kids. Someone like me, how do I compete? How do you compete, so, my whole thing is, I do programs like this to try to compete. (…) And my edge is going to be experience and exposure and professionalism. (Interview with Martyn, 2015)

He was clearly aware that being able to have the proof for his internship in Washington, a letter of recommendation and a certificate from the WSP would be the proof he needed for his mobility experience in a place of power which would then help him to further his career. Moreover, as this quote suggests while claiming to not be as clever as some of his competition, he indirectly saw himself in competition with students who went to the more elite universities in the US. His solution to this competition was participation in programmes like the WSP in order to become more experienced and professional. Thus, he concluded that a time in Washington, at a University and in an internship was a way to replace studying at an elite university. As Perlin suggests, there might be a case to be made for Martyn's reasoning as in recent years,

> Dozens upon dozens of schools have set up their own Beltway operations in the last few decades, largely to position their students for the internship feeding frenzy. Among the most prominent are programs run by Cornell, Claremont McKenna, the University of California system, Syracuse, Boston University, Harvard Law School and Stanford, but there are many more. Between these university beachheads, the massive non-profit internship centres, and personal connections, young people on their own stand little chance of landing a well-placed internship in DC, if they can even afford it to begin with – given an estimated cost of living around $1,500 per month – –on a responsible student's budget. (Perlin, 2011, p. 111)

As some government departments increasingly source out their internship recruitment to programmes at the Washington Center (Perlin, 2011, p. 109), individual internship opportunities become sparse and students are indirectly forced to rely on study-internship programmes in Washington to find internships. Ploner (2015, p. 2), while acknowledging the number of cosmopolitan study and learning opportunities that have been developed in the global knowledge economy, also notes that 'it is also characterised by uneven affordances and power relations which marginalise those who are "immobile" due to social, financial or political reasons'. Frändberg (2014, p. 148) further explains that there is a negative side to the increasing number of mobility opportunities for students:

> … the "freedom to explore" has another side, which is mobility as a strategy for handling increasing labour market insecurity and perhaps also for fulfilling expectations of becoming a (geographically) flexible adult. In certain social groups, transnational mobility competence is increasingly seen as a precondition for employability …

One danger of mobility programmes is that as there are many families and students who are not able to afford these programmes and are not able to invest in their children's cultural and human capital, such programmes will lead to further socio-economic divisions.

Brad, 31, a WSP alumnus, elaborated on the competitive nature of social relationships in Washington and how even his private life was often shaped by networking, self-affirmation and competition:

> it is very elite-like transaction driven city, where everyone you meet: the first … questions are like: What do you do? i.e. How important are you? Where do you live? i.e. How much money do you have? Yeah, where do you live? What do you do? And where do you live? And those are just, it is kind of an instant sizing up or putting in somebody into like a certain bucket. And then the third question is basically how valuable are you to me? They don't ask that directly but that is at the back of their head. (Interview with Brad, 31, American WSP Alumnus, 2015)

As Brad also added,

> in DC if you tell a good story and you are compelling, I think that is most important (…) people don't buy what you do, they buy why you do it (…) [t]hey want to know your motivation and they want to know that you are like succeeding on your things and can tell a good story. (Interview with Brad, 31, American WSP Alumnus, 2015)

A part of selling of your own story is to identify as a global citizen, especially for young people this seems essential to obtain a job in a transnational company or an NGO (Jones, 2011, p. 532). The WSPs themselves also tell a story, and convey a feeling of being part of the inner circle in the DC, the circle of 'decision-makers'. As George summarises and other participants also suggested, a feeling of being privileged seems to be common among WSP participants, due to the exclusive nature of the classroom activities, and trips that the WSP classes do in Washington:

> I mean sometimes, you know, you get into the World Bank you are talking to the communications director and you are like 'wow, this is a really important person', no one's getting to like listen to this, really. There is not many people that get to listen to stuff like this. (…) And seeing every major organization over the course of a semester, where you like, there's not, you mean if you haven't seen the rest of the city you will still kind of figure it out, because you've been exposed to everything. So you get an education in Washington DC if anything. They are just not here to teach blunt material and textbook stuff. Like this is how the city operates and here is how to work it, if you want it you can have it here. It's huge, it's just a taste. (George, American, 22)

As this quote reveals 'the education in Washington DC' perfectly summarises the acquisition of very specific cultural capital that other competitors on the global job market are lacking and which students want to acquire. As a Higher Executive of the WSP stated:

> We've had lot of students who say that they're glad they learned when they were with us that this thing was not for them. And, many times they also identify what is for them; because they've been exposed to things they have never thought about or heard about before. And then they follow up with those people, and network with them, either to get another internship for the following year or for the summer, or just to get in touch with them to learn what kind of classes they should take to be able to start their careers; so in that sense I think you can say that people can jumpstart their careers. (Interview with a Higher Executive of the WSP, 2015)

As can be identified in this quotation, the 'education in Washington DC' has many components, it is about self-search and questions of identity and possible careers for young people, as well as about networking, whether on the internship sites or among the classmates within the programmes or also with guest speakers. All of this showcases the ways in which these programmes are orchestrating the various resources of Washington, DC and put the participants in a position to reap those benefits. Through these actions, the WSPs contribute to the reproduction of elites within in the city of Washington, DC. This is further exemplified below in the following section that discusses further the mobilities of students, their reflections on employability and the influence of the transience of the city of Washington, DC.

Mobilities and transience

The notion of mobility as a means to prepare for the labour market and increase employability options might also be seen as impacting mobility decision-making and restricting the freedom of choice. As the example of Martyn shows, these programmes are utilised as a means of increasing student's human capital value and employability. Especially with a closer look at the relation of human capital theory and internships, Perlin states that 'students and their families may feel compelled to invest heavily in education and skills-building' (2011, p. 129), but the idea is that they will be rewarded for this investment. Furthermore, he adds, 'Economists assume that this "college bonus" (and any "internship bonus" that might exist) accurately reflects the results of human capital investment: better work and greater productivity, which employers are quick to reward' (Perlin, 2011, pp. 129–130). The idea of the 'mobility burden' (Shove, 2002), the implicit necessity to be mobile, becomes important here as increasingly students like Martyn feel they are expected to join such global study-internship programmes in order to become valued members of society in competition with elites. Conversely, for other young people, the concept of home and the local may regain popularity as the pressures to be mobile become too much or may not fit into their value systems.

As many young people try to go to Washington for an internship or for undergraduate or graduate studies, in the interviews conducted, the city was often described as an extremely transient place, as people tended to live there for a couple of years or months.

> What I find difficult about Washington is that there are many people mmm that move to DC after um, after finishing their Masters or maybe for their Masters and then they stay for a few years and then they move on. So in a way it's not a place where you have like many real neighbourhoods. I feel and it's not a place um where really people um, um stay to live. They come for a career and they might leave again. (Interview with Aaron, 2015)

Here, Aaron suggests that many people do not associate and measure Washington that much by terms of quality of living but rather in terms of usability for their careers. Aaron highlights that DC is not a place where people 'stay to live'. He showcased a perspective that emphasises the value of career aspects in his value system. Another participant, Justine, who was working as an environmental lawyer, provided a very similar perspective, as she asserted:

> another thing that I think is unique about DC: it is very transient city, which is why rent is so high (Laughing), it is a very big renters market of people who are just coming for a few years, like jumpstart their career to like get a certain type of experience. (Interview with Justine, 2016)

These statements confirm Frändberg's argument that, 'at least for large groups in the world's richer countries, long-distance temporary moves have become a significant part of the transition to adulthood', especially as they help young people in 'exploring future social and professional opportunities as well as part of the "project of the self"' and may substantially impact their future mobilities (Frändberg, 2014, p. 149).

George also elaborated on the culturally attractive factors of Washington and the cosmopolitan atmosphere of the city and social interactions.

> Yeah, socially it's great. Everyone that you engage with, I mean there is like a big , you know the nightlife scene is totally thriving here, there is a lot of young people, you know most, I don't know what the stat was, some ridiculous stat about everyone living here from like twenty to thirty years old, it's like a place for young professionals, so. People are always out and engaging in the city, with events and music, you know going out to the bars and the restaurants. When you do engage it is, it is (…) when you engage with them, you like it is really stimulating. Everyone's very smart, everyone kind of has a role, if you are in the city, you are kind of, you are not here to you know just work and live a normal job like there is something that you are going to be doing in the city because, just the chance you have of meeting someone that is doing something cool in the city is so high, so you can always have a conversation about what they are doing, what you are doing, somehow it relates and you have a great rich conversation, often intelligent and it's fun. (Interview with George, 2015)

What is significant about George's remarks is how interwoven the cultural aspects that he highlights are with career aspects of getting to know people and networking. He notes the chances of meeting someone 'cool' is very high and he finds engaging with the city and its people stimulating. This helps to depict a culture of constant networking that is present in the leisure nightlife of the city. It also re-enforces the image that 'outsiders' might have of life within the beltway, as it describes a bubble in which the inhabitants of Washington, DC take themselves very seriously and have very political debates that might seem strange to other Americans. Moreover, it is interesting that George highlighted the city as very young and fun some, an image that might in some ways contradict images of the 'old elites' that run the US and the city. Also, the emphasis on how many people might possibly be interesting or relevant to George also showcases the transience of the city as well as how fluent and short-lived personal relationships are in George's experience. In addition, to these factors, these remarks show that career aspects are a dominant theme in the WSP participant's mobility decision-making, but once the students arrive in Washington, the factor of personal development, opportunity for individualisation and participating in global citizenship lifestyles in Washington play a significant role in this form of student mobility and experiencing Washington, DC. While George, who was still very new to the city was excited by these aspects, other interviewees who had lived in Washington for a couple of years found the aspect of constant networking very exhausting and tiring. Albert, 24, a German Alumnus of a WSP also asserted that being in DC politicised him and he started to develop new ideas for future career paths due to the city. Albert interned for a medium-sized international NGO and stated that he,

> got way more enthusiastic about the topics and [when we went] to all the NGOs and public agencies and whatever, and it actually kind of triggered that I got more interested in this topic. Again. And that was also kind of very informative. (..) Also for my later kind of career plans. That was the big thing about the Washington Semester Program. It helped me a lot to find out basically what I can actually achieve, especially talking to people. (Interview with Albert, 2016)

Just the exposure to the city of DC, especially within these WSPs, seems to have the capacity to shape their career and mobility paths. As Albert's quote reveals, self-discovery, individualisation and exposure to the 'NGOsphere' of Washington were realised through the participation in a WSP.

Alice, 21, an American student from California, explained that there were a number of reasons that made her come to Washington for a WSP:

> I studied abroad a year ago, and I was studying in El Salvador, when I returned I was in this kind of middle space where I had a very positive experience abroad, I spent a lot of time in the community of women and children, very impoverished community. (…) Back at school, and so I came back and I knew that I wanted to be a child studies major (…) but as far as career, I knew that it was never too early to start thinking about that, and feeling that I knew that exposure, I heard about the WSP, my school has a partnership with AU, which makes it really easy to come here (…) wanting to

explore specifically the area of policy, but at the same time had that component of (unintelligible) which is why I had an internship at a non-profit. And so having heard how the program really did a good job of combining the two, and giving us exposure to, being in DC, you see the policies, you see the politics, how that plays in, in with communities, and then how those communities, how non-profits fill in the gap, so that's what I was really looking in and for and I wasn't sure I was going to get that elsewhere. So it was really about coming to DC for me. (Interview with Alice, 2015)

The direct exchange agreement between her home school and American University seemed to her as a simple way to gain this experience, as she only needed to pay the regular tuition of her home school. Moreover, she wanted to gain more experience in the field of community work. It should be mentioned that the vocabulary that Alice used was very specific and seemed to reflect the language that is used in the programme's brochures, as well as her class teacher of the sustainable development class and in general in Washington, DC.

Alice had been to Washington before participation in the programme because her sister was living in the city, and she stated that her 'sister did a lovely job of showing me around DC and from that trip, I knew I wanted to see more of DC, I knew I wanted to come back' (Interview with Alice, 2015). There were also other participants who highlighted previous trips to family members or friends in Washington as well as high school trips to Washington, DC. VFR (Visiting Friends and Relatives) connections and associations with Washington, DC were a key factor for their decision-making (see Boyne, Carswell, & Hall, 2002).

Conversely, however, Nathalie, who was nominated for the programme, asserted that there were other more cultural reasons for coming to Washington, DC:

There is a great live music scene in DC, there is poetry which I really like. (…) One big thing, a factor when I am deciding to move somewhere is how easy it is to get around on public transportation, because at that time; although I had my driver's license I did not have a car; so I knew that I would be able to get around just fine. And I knew that my cousin would still be there, so I would have someone that I; I at least knew one person; I did not know anyone who was going to participate in the program but I know if I wanted an out I had family in the city that I could go and hang out with; so I think that made the decision a little bit easier, too. (Interview with Nathalie, 2015)

So, in her case, there were many factors, the nomination for the programme that made her aware of the programme in Washington, as well as the general possibility to go and take part in a programme at a different university. As her family received no tertiary education in the US, she did not have the cultural capital and required knowledge about these opportunities. An attraction to the cultural possibilities in Washington as well as a more practical mundane mobility reason, the accessibility of Washington, DC via public transport, because she did not own a car all factored into Nathalie's decision of taking part in this programme in Washington. Also in this case, there was a family member who was living in the city and alleviated her decision to move into a different city. Nathalie's case exemplifies the multitude of factors that play a role in educational mobility decisions.

Conclusions

Washington, DC is an extremely transient city but with the election of President Trump, it remains to be seen whether DC will remain a focus for international students. In this paper, we have explored how Washington, DC has become a hub for student's seeking to become part of the global elite through participation in study internships which are promoted as enabling them to become global political citizens. In particular, we have shown that apart from the unique study and networking opportunities of participation in one of these programmes, increasing one's cosmopolitanism and global citizenship plays a significant part in the promotion of these programmes.

Moreover, career aspects are a dominant theme in the participant's mobility decision-making, but once the students arrive in Washington, the factor of personal development, opportunity for individualisation and participating in global citizenship lifestyles in Washington play a significant role in this form of student mobility and experiencing Washington, DC. While the study-internship programmes emphasise that students will experience more of Washington, DC than a tourist, this has conversely helped to re-create Washington, DC as an increasingly transient city experienced by both students and tourists as a place that you would not 'stay to live' long term. This highlights the mobilities of both place and people, as Washington, DC has become a city of mobile global citizenship where the inequalities of access to power are often hidden within networks of cultural capital and cosmopolitanism.

Disclosure statement

No potential conflict of interest was reported by the authors.

ORCID

Felix Schubert http://orcid.org/0000-0002-6671-2225
Kevin Hannam http://orcid.org/0000-0001-6034-3521

References

Adey, P. (2010). *Mobility*. London: Routledge.
Adey, P., Bissell, D., Hannam, K., Merriman, P., & Sheller, M. (2013). *The routledge handbook of mobilities*. London: Routledge.
Allen-Robertson, J., & Beer, D. (2010). Mobile ideas: Tracking a concept through time and space. *Mobilities*, 5(4), 529–545.
American University. (2010). Connect where it counts. Retrieved November 1, 2016, from https://issuu.com/raihansdg/docs/connect
American University. (2014). Washington Semester Program live, learn, know. Retrieved November 1, 2016, from http://issuu.com/au_spexs/docs/wsp_brochure_2014
American University. (2016). Admission requirements for international students. Retrieved November 5, 2016, from http://www.american.edu/spexs/washingtonsemester/Admission-Requirements-For-International-Students2.cfm
Benson, M., O'Reilly, K., & Kershen, A. J. (Eds.). (2009). *Lifestyle migration: Expectations, aspirations and experiences*. Farnham: Ashgate.
Boyne, S., Carswell, F., & Hall, D. (2002). Reconceptualising VFR tourism. In C. M. Hall, & A. Williams (Eds.), *Tourism and migration* (pp. 241–256). Dordrecht: Springer.
Butcher, J., & Smith, P. (2015). *Volunteer tourism: The lifestyle politics of international development*. London: Routledge.
Chandler, D. (2009). *Hollow hegemony*. London: Pluto Press.
Cohen, S. A. (2010). Personal identity (de)formation among lifestyle travellers: A double-edged sword. *Leisure Studies*, 29, 289–301.
Cohen, S. A., Duncan, T., & Thulemark, M. (2015). Lifestyle mobilities: The crossroads of travel, leisure and migration. *Mobilities*, 10, 155–172.
C-SPAN. (2004). John McCaslin: Inside the Beltway. *Booknotes*. Retrieved November 5, 2016, from http://www.book-notes.org/Watch/181421-1/John+McCaslin.aspx
DC Internships. (2016). About us. Retrieved September 22, 2016, from: http://www.dcinternships.org/about-us/
Erickson, A. (2012). Washington, D.C.'s tourist trap. Retrieved November 8, 2016 from: http://www.citylab.com/politics/2012/04/washington-dcs-tourist-trap/1704/
Frändberg, L. (2014). Temporary transnational youth migration and its mobility links. *Mobilities*, 9(1), 146–164.
Hannam, K., & Guereno-Omil, B. (2015). Educational mobilities: Mobile students, mobile knowledge. In D. Airey, D. Dredge, & M. Gross (Eds.), *The routledge handbook of tourism and hospitality education* (pp. 143–154). London: Routledge.
Hannam, K., & Knox, D. (2005). Discourse analysis in tourism research a critical perspective. *Tourism Recreation Research*, 30(2), 23–30.
Hannam, K., & Knox, D. (2010). *Understanding tourism: A critical introduction*. London: Sage.
Hannam, K., Sheller, M., & Urry, J. (2006). Editorial: Mobilities, immobilities and moorings. *Mobilities*, 1(1), 1–22.
Jackson, J. (2015). The consequences of gentrification for racial change in Washington, DC. *Housing Policy Debate*, 25(2), 353–373.
Johnson, J. (2010). More would-be interns paying thousands to land a coveted spot. *The Washington Post*. Retrieved from http://www.washingtonpost.com/wp-dyn/content/article/2010/08/29/AR2010082903743.html
Jones, A. (2011). Theorising international youth volunteering: Training for global (corporate) work? *Transactions of the Institute of British Geographers*, 36, 530–544.
Knox, L. (1991). The restless urban landscape: Economic and sociocultural change and the transformation of metropolitan Washington, DC. *Annals of the Association of American Geographers*, 81(2), 181–209.
Lam, T., & Ching, L. (2007). An exploratory study of an internship program: The case of Hong Kong students. *International Journal of Hospitality Management*, 26, 336–351.
Maher, J. T. (2015). The capital of diversity: Neoliberal development and the discourse of difference in Washington, DC. *Antipode*, 47(4), 980–998.
Marchio, N., & Berube, A. (2015). *Benchmarking greater Washington's global reach: The national capital region in the world economy*. Washington, DC: The Brookings Institution.
McCaslin, J. (2004). *Inside the beltway: Offbeat stories, scoops, and shenanigans from around the nation's capital*. Nashville, TN: WND Books.
Mostafanezhad, M. (2013). The geography of compassion in volunteer tourism. *Tourism Geographies*, 15(2), 318–337.
Mostafanezhad, M. (2014). Volunteer tourism and the popular humanitarian gaze. *Geoforum*, 54, 111–118.
Perlin, R. (2011). *Intern nation: How to earn nothing and learn little in the brave new economy*. Brooklyn, NY: Verso Books.
Ploner, J. (2015). Resilience, moorings and international student mobilities – Exploring biographical narratives of social science students in the UK. *Mobilities*, doi.org/10.1080/17450101.2015.1087761.
Politico. (2009). D.C. interns by the numbers. Retrieved August 26, 2016, from http://www.politico.com/story/2009/07/dc-interns-by-the-numbers-024883
Reuters. (2015). Washington, D.C. set tourism record in 2014 as foreign visits surged. Retrieved November 08, 2016 from: http://www.reuters.com/article/us-usa-districtofcolumbia-tourism-idUSKCN0QM21720150817
Rickly, J. M. (2014). Lifestyle mobilities: A politics of lifestyle rock climbing. *Mobilities*, 11, 243–263.
Root, E., & Ngampornchai, A. (2013). 'I came back as a new human being' Student descriptions of intercultural competence acquired through education abroad experiences. *Journal of Studies in International Education*, 17(5), 513–532.
Sheller, M., & Urry, J. (Eds.). (2004). *Tourism mobilities*. London: Routledge.
Shove, E. (2002). *Rushing around: Coordination, mobility and inequality*. Draft paper for the Mobile Network meeting, October 2002. Lancaster. Retrieved from http://www.lancaster.ac.uk/staff/shove/choreography/rushingaround.pdf
Standish, A. (2012). *The false promise of global learning – Why education needs boundaries*. London: Bloomsbury Open.

Stanford in Washington. (2016). Stanford in Washington Program. Retrieved September 22, 2016, from https://siw.stanford.edu

The Osgood Center. (2016). Washington Experience – Leadership and professional development program. Retrieved November 7, 2016, from http://www.osgoodcenter.org/washington-experience.php

Thorpe, H. (2012). Transnational mobilties in snowboarding culture: Travel, tourism and lifestyle migration. *Mobilities*, 7, 317–345.

Trujilo, J. R., & Parilla, J. (2016). Redefining global cities. *Global Cities Initiative*. Retrieved November 5, 2016, from https://www.brookings.edu/research/redefining-global-cities/

UCDC. (2016). Who we are. Retrieved September 22, 2016, from https://www.ucdc.edu/who-we-are/programs-dc

Washington Center. (2015). *Experience transforms*. Washington, DC: The Washington Center for Internships and Academic Seminars.

Washington Center. (2016). Accomplishments. Retrieved September 22, 2016, from http://www.twc.edu/about/accomplishments

Williams, A. (2006). Lost in translation? International migration, learning and knowledge. *Progress in Human Geography*, 30 (5), 588–607.

Producing science and global citizenship? Volunteer tourism and conservation in Belize

Noella J. Gray, Alexandra Meeker, Sarah Ravensbergen, Amy Kipp and Jocelyn Faulkner

ABSTRACT
Many volunteer tourism programmes in the Global South involve volunteers in the collection of biological data for the purpose of environmental conservation and monitoring. By participating as lay people in the collection of scientific data, volunteer tourists are similar to citizen scientists in the Global North. Both activities are part of a neoliberal science and conservation regime. Building on insights from citizen science and analyses of global citizenship in tourism, this paper argues that global citizenship and scientific knowledge are co-produced through volunteer tourism. Drawing on the results of a case study of a volunteer tourism and marine conservation programme in Belize, including 48 interviews and 4 focus groups with a range of actors (foreign volunteers, staff of government agencies and non-governmental organisations, and local residents), this paper illustrates both the opportunities and limitations for producing critical global citizenship and democratised science through such programmes.

Introduction

Volunteer tourism is a rapidly growing type of alternative tourism, in which tourists pay to undertake volunteer work as part of their vacation. Both the practice of volunteer tourism and scholarship focused on it have grown exponentially in the past 20 years, with early scholarship focused primarily on motivations of volunteers, the role of volunteer tourism organisations, and to a lesser extent, the experiences of host communities (Wearing & McGehee, 2013). More recently, scholars have begun to examine volunteer tourism in relation to broader political-economic processes and theoretical approaches, including neoliberal development and governmentality (Mostafanezhad, 2014; Sin, Oakes, & Mostafanezhad, 2015; Vrasti & Montsion, 2014), geographies of care and compassion (Mostafanezhad, 2013; Sin, 2010), and global citizenship (Butcher & Smith, 2015; Daly, 2013; Lorimer, 2010; Lyons, Hanley, Wearing, & Neil, 2012); this paper contributes to the latter strand of scholarship. Citizenship is a complex, widely used term that intersects with tourism in numerous ways (Bianchi & Stephenson, 2014). The term 'global citizenship' refers to the rights and responsibilities individuals have in relation to all humans (and non-human nature), not only those within a particular nation; volunteer tourism is often discussed as a means for individuals to exercise their responsibilities to act ethically toward a 'global community' (Bianchi & Stephenson, 2014; Butcher & Smith, 2015).

Volunteer tourism is a 'macro-niche' industry (Stainton, 2016), including a wide range of activities such as volunteering on farms (Yamamoto & Engelsted, 2014), in orphanages (Rogerson & Slater, 2014), and supporting wildlife conservation (Lorimer, 2009). We focus on the latter form of volunteer tourism, sometimes referred to as conservation tourism (Cousins, Evans, & Sadler, 2009), conservation voluntourism (Brondo, 2015), volunteer ecotourism (Gray & Campbell, 2007), or research ecotourism (Clifton & Benson, 2006; Galley & Clifton, 2004). International conservation volunteer programmes are prevalent across the Global South (Lorimer, 2009). While projects include a variety of activities designed to support conservation, many of them involve volunteers in environmental monitoring and scientific data collection. For example, volunteers may collect data on sea turtles (Gray & Campbell, 2007), macaws (Brightsmith, Stronza, & Holle, 2008), or as in this case, fish and benthic species associated with coral reef habitat.

Whenever they participate in the collection of scientific data, volunteer tourists are similar to citizen scientists. Citizen science refers to any project in which non-professionals (citizens) engage in scientific research, by forming research questions, collecting data, and/or analysing results (Dickinson et al., 2012; Miller-Rushing, Primack, & Bonney, 2012). In a recent call for tourism research that is more hopeful and less dystopian, Brosnan, Filep, and Rock (2015) specifically call for attention to the insights of citizen science and note that 'volunteer tourists are increasingly engaging in citizen science projects' (p. 97). They suggest that 'research projects could explore the role of citizen science tourism experiences in promoting environmental and humanitarian awareness, compassion and activism' (Brosnan et al., 2015, p. 98). It is here that attention to recent scholarship regarding global citizenship and volunteer tourism is relevant. However, rather than assume that volunteer tourism and citizen science are inherently more inclusive, democratised forms of knowledge production (Brosnan et al., 2015), we propose this should instead be the focus of empirical research.

In this paper, we argue that global citizenship and knowledge are co-produced through volunteer tourism, offering both limitations and opportunities for the development of more critical forms of global citizenship. In the sections that follow, we first consider the relationship between global citizenship and volunteer tourism, highlighting features of 'critical' global citizenship as well as the need for attention to the role of knowledge production in mediating this relationship. Next, we discuss the similarities and differences between volunteer tourism and citizen science, identifying lessons from citizen science that are relevant for assessing volunteer tourism. We then discuss our case study of volunteer tourism in Belize, reviewing the details of the case, our methods, and the results of our study. We interpret our results in relation to critical global citizenship and lessons from citizen science. Finally, we conclude by discussing both the limitations and opportunities inherent in volunteer tourism as a means for co-producing knowledge for conservation and critical global citizenship.

Volunteer tourism, global citizenship, and science

Politicians, educators, and volunteer tourism organisations all widely promote volunteer tourism as a means of developing global citizenship (Butcher & Smith, 2010; Lyons et al., 2012). However, the concept of global citizenship is often ill-defined. Cameron (2013) explains that the term's popularity is in part due to its user friendliness, as global citizenship discourse is used to promote both idealistic ideas of travel as well as neoliberal ideals of workforce competitiveness. To understand global citizenship, many scholars have turned to the related concept of cosmopolitanism, which emphasises a concern for the wellbeing of humankind regardless of borders, and celebrates diversity and human rights (Lyons et al., 2012).

Drawing on the concept of cosmopolitanism, scholars focused on education and experiential learning have distinguished between soft or thin forms of global citizenship and more critical or thick conceptions (Andreotti, 2006; Cameron, 2013). Soft global citizens assume that problems in the Global South result from a lack of access to resources such as technology and education, thereby failing to acknowledge the political and historical processes that reinforce inequality and reproducing the colonial idea that the West can save the 'rest' from their poverty, simply by imparting their knowledge (Daly, 2013). In contrast, critical global citizens attempt to understand and change the structures and systems at the root of inequality and poverty and examine the role they play in these systems (Cameron, 2013; Daly, 2013). Critical global citizenship thus draws on the ethics of cosmopolitanism, promoting the responsibility of the individual in actively working to disrupt the social structures that maintain a global hierarchy (Andreotti, 2006; Cameron, 2013). As detailed by Andreotti (2006), there are several features of critical global citizenship (versus soft global citizenship) worth noting in relation to knowledge production specifically. First, soft global citizens feel a responsibility to *teach* the 'other', while critical global citizens are responsible to *learn with* the 'other'. Second, soft global citizens donate their expertise and resources whereas critical global citizens analyse their own position and participate in changing assumptions and identities. Third, soft global citizens assume a universal understanding of how everyone should live (or in this case, learn/know), while critical global citizens engage in reflexive dialogue and ethical relations with difference (including different ways of knowing). At the same time, critical citizenship cannot reject universalism (and embrace relativism) to the extent that it inhibits political action (Butcher & Smith, 2015; Cameron, 2013).

There is substantial debate as to whether and how volunteer tourism can foster global citizenship. Volunteer tourism may provide individuals from the Global North with the opportunity for more genuine exchange with individuals in the Global South (Raymond & Hall, 2008). These experiences can heighten the cross-cultural understanding and civic engagement of volunteers, both at home and abroad (Conran, 2011). While this falls short of constituting critical global citizenship (Conran, 2011), it is an important step in reframing the relationship between the North and the South and building broader networks of solidarity (McGehee, 2012). However, cross-cultural understanding is not an inherent outcome of volunteer tourism. Rather than challenging assumptions of cultural superiority, patterns of privilege, and global inequalities, volunteer tourism may reinforce them (Borland & Adams, 2013; Lyons et al., 2012; Zeddies & Millei, 2015). Moreover, volunteers attempt to 'make a difference' while building their resumes (Simpson, 2005), constructing themselves as 'competitive, entrepreneurial, market-based, individualized actors *and* caring, responsible, active, global citizens' (Sin et al., 2015, p. 122). Whether critical global citizenship can be built through volunteer tourism, in a way that is not reducible to neoliberal modes of 'consumer citizenship', or, conspicuous (ethical) consumption, is doubtful (Bianchi & Stephenson, 2014).

A related concern is whether and how global citizenship extends to *all* individuals involved in volunteer tourism, not just volunteers. The volunteer tourism industry is predicated on the notion that the 'better off' should help the 'worse off', limiting the possibility of a reciprocal relationship in which both actors benefit from the interaction (Lyons et al., 2012). Volunteer tourists are depicted as global citizens who have agency by traveling to 'help the less fortunate' (Daly, 2013, p. 71). By extension, individuals in communities of the Global South lack agency, limiting their claim to global citizenship and rendering them 'second class citizens' in the world system (Zeddies & Millei, 2015, p. 103).

While the literature on volunteer tourism has debated whether and to what extent volunteer tourism promotes (critical) global citizenship, it has not yet considered how this is mediated by and through knowledge production activities. It is important to consider whether participation in science facilitates models of citizenship founded on social solidarity or political marginalisation (Leach & Scoones, 2005). In conservation volunteering, volunteers are aligned with a discourse of global *environmental* citizenship, demonstrating care for a global nature that transcends national borders (Bianchi & Stephenson, 2014). This discourse intersects not only with ideas of 'global' space and neoliberal modes of delivery, but also relies on 'science as the guiding logic' (Lorimer, 2010, p. 313). How does the integration of science affect the relationship between volunteer tourism and global citizenship? In what ways does scientific knowledge production enable or constrain more critical forms of global citizenship? We turn to the literature on citizen science in order to identify relevant lessons for volunteer tourism.

Comparing volunteer tourism and citizen science

There are important similarities between volunteer tourism and citizen science. First, both types of projects support knowledge production by providing laypeople with the opportunity to participate in the collection of biophysical data. It would be difficult to collect comparable data sets without volunteer labour; this free labour enables the collection of long-term data across large areas, a feature of particular value for environmental monitoring and conservation-related research (Brightsmith et al., 2008; Dickinson et al., 2012). Many citizen science projects contribute to scientific discoveries (Follett & Strezov, 2015), while individual case studies suggest that some volunteer tourism projects are also contributing to the advance of science through publications (Brightsmith et al., 2008). However, not all projects will (or should) lead to academic publications. Citizen science projects may also have 'action' or conservation goals (Follett & Strezov, 2015), where providing relevant data to local groups working on conservation and management issues is the objective (Jansujwicz, Calhoun, & Lilieholm, 2013). This leads to a second benefit, namely social outcomes for communities. While the primary objectives of citizen science projects are scientific, they may also achieve social goals, such as supporting sustainable livelihoods or documenting negative environmental or health impacts (Bonney et al., 2014). Volunteer tourism can also (though does not always or necessarily) provide important social and economic benefits for local communities (Wearing & McGehee, 2013). Third, both activities provide benefits for participants. Benefits for volunteer tourists include skill development (Lyons et al., 2012; Wearing & McGehee, 2013), personal growth and increase in confidence (Alexander, 2012; Barbieri, Santos, & Katsube, 2012; Coghlan & Gooch, 2011), and increased environmental consciousness (Weaver, 2015). Similarly, citizen scientists report acquisition of skills, knowledge, and scientific literacy (Donnelly, Crowe, Regan, Begley, & Caffarra, 2014; Johnson et al., 2014), and empowerment to participate more broadly as advocates and opinion leaders (Jansujwicz et al., 2013; Johnson et al., 2014). Finally, both volunteer tourism and citizen science are implicated in broader neoliberal policy shifts, which have reduced state funding for science and conservation and emphasised the role of markets and entrepreneurial individuals in advancing social goals (Brondo, 2015; Cousins et al., 2009; Lyons et al., 2012). They are both products of what Rebecca Lave calls the 'neoliberal science regime' (Lave, 2012).

Although there are several important similarities between citizen science and volunteer tourism, they are distinct in two important respects: cost and location. Citizen scientists may incur some nominal costs to support their participation, but typically there is no fee associated with these projects. In principle, citizen science projects should be open to all citizens (Stevens et al., 2014). In contrast, volunteer tourists pay a fee to participate in conservation projects, typically in the range of £225–£750 per week, not including travel costs (Cousins, 2007; Lorimer, 2009). There is thus an inherent elitism to volunteer tourism, as the cost of such placements render them inaccessible to those lacking disposable income (Lyons et al., 2012). In practice, citizen science may also be less accessible to some individuals, particularly lower-income and less formally educated individuals (Edwards, 2014; Pandya, 2012). However, while there are barriers to engaging marginalised communities in citizen science projects, these can be overcome (Stevens et al., 2014). Location also differentiates citizen science from volunteer tourism. Citizen scientists typically collect data 'in their own backyards', (sometimes literally, as in the Great Backyard Bird Count), and often focus on local or place-based topics and problems (Miller-Rushing et al., 2012). Citizen science tends to engage people with 'familiar' things they already know or encounter regularly (birders who count birds), whereas volunteer tourism attracts people who want to experience 'exotic' ecosystems and charismatic species (Lorimer, 2009) and invokes a 'global spatial imaginary' (Lorimer, 2010). Citizen science happens at home, for free; volunteer tourism happens anywhere other than home, for a price.

Thus far, the term citizen science has been used as it is discussed in ecology and natural science literature focused on public participation in large-scale data collection (e.g. Dickinson et al., 2012; Follett & Strezov, 2015). However, in the science and technology studies (STS) literature, citizen science refers to the democratisation of science through the integration of lay concerns and marginalised communities into scientific discourse and practice, often aligned with goals related to social and environmental justice (Irwin, 1995; Lave, 2012; Shirk et al., 2012). The STS literature thus draws attention to the relationship between scientific knowledge production, empowerment, and citizenship. Jasanoff (2004), for example, outlines how citizens both produce and consume knowledge as a means of political participation, while Leach, Scoones, and Wynne (2005, p. 13) note that 'scientific engagement makes citizens' through the complex interplay of scientific knowledge practices, human relations and identities, and politicised institutions.

Building on the citizen science literature reviewed above, in conjunction with a review of the literature on public participation in scientific research (i.e. citizen science) (Shirk et al., 2012), two lessons emerge. First, in addition to scientific goals and outcomes, projects should also be attentive to 'action' goals and outcomes that may include supporting conservation monitoring, policy and management, increasing citizens' skills and knowledge, and increasing trust among stakeholders in order to support collaborative management of social-ecological systems. Second, in order to truly democratise science – to integrate marginalised communities and their concerns into scientific practice – careful attention must be paid to the degree and quality of participation. This includes consideration for the extent to which citizens are involved in all aspects of the scientific research, for mechanisms to overcome barriers to participation, as well as for the context for participation in science along dimensions such as trust, credibility, fairness, relevance, and agency (Shirk et al., 2012). We contend that these lessons from citizen science are relevant for understanding whether and how the integration of science affects the relationship between conservation volunteer tourism and (critical) global citizenship. We turn to our case study of conservation volunteering in Belize in order to illustrate this.

Methods and study site

This paper is based on a case study of the volunteer tourism programme run by Blue Ventures (BV) in Belize. Tourism is one of the largest economic sectors in Belize, contributing 38.6% of the country's GDP in 2015 and 35% of total employment (WTTC, 2016). One of the main tourist attractions is the marine environment. Belize is home to a large portion of the Mesoamerican Barrier Reef System and has established 15 marine protected areas to support its conservation (Cho, 2005; Government of Belize, 2015). Non-state actors, primarily non-governmental organisations (NGOs), have played a critical role in conservation and ecotourism activities in Belize (Medina, 2015), including in marine conservation.

BV, a UK-based organisation, aims to go 'beyond conservation', providing meaningful and lasting benefits for their partner communities by diversifying and strengthening local economies and empowering local people to protect the marine environment (BV, 2016). BV consists of two separate entities, BV Expeditions and BV Conservation. BV Expeditions, a registered company in Scotland, operates the volunteer 'marine conservation expeditions' and generates revenue through the collection of fees from volunteers. Instead of generating funding for shareholders, these profits are reinvested into BV's social enterprise, BV Conservation. In addition to funding generated by volunteer expeditions, BV Conservation also receives funding from corporations, foundations, and governments.

The typical BV volunteer programme in Belize is six weeks in duration; these trips are offered throughout the year, in addition to shorter trips at specific times. While BV volunteers have diverse backgrounds, the majority are 'mid-career professionals taking career breaks, gap year and university students and retirees' (Blue Ventures, 2015), with an average age of 28 (Blue

Ventures, 2017). Volunteers spend the first and last week of their trip in the community of Sarteneja, staying with a local family through the Sarteneja Homestay Group, while the middle four weeks are spent at the BV dive camp within the Bacalar Chico Marine Reserve and National Park (BCMRNP) (see Figure 1). It is important to note that BCMRNP is not adjacent to the community, nor is it heavily used by tourists. The BV dive camp within BCMRNP is remote and offers only basic amenities.

During the first week, volunteers acquire the knowledge necessary to fulfil their role as data collector, listening to lectures and studying for the tests they must pass related to fish or benthic species identification. During their four weeks at the dive camp, volunteers undergo further training and testing, followed by participation in standardised data collection procedures while scuba diving. The final week is spent back in the community of Sarteneja, participating in outreach and other community-based activities led by BV. As of 2016, the cost for a six-week expedition (not including international airfare) was approximately £2600–2900. The data collected by volunteers inform the annual reports that BV compiles (e.g. Chapman, 2014), which are shared with the Fisheries Department, the government agency responsible for management of the marine reserve portion of BCMRNP and for permitting BV's work there. Volunteer-collected data also contribute to the information that BV shares with national and international conservation networks and initiatives.

Sarteneja is a community of approximately 1824 people in the north of Belize (Statistical Institute of Belize, 2013). While the village was once the site of an ancient Mayan settlement, its current population is predominantly Spanish-speaking Mestizo. After a severe hurricane and period of drought in the 1950s, Sarteneja was forced to transition away from agricultural livelihoods, to fishing (SACD, 2009). Although many coastal communities in Belize have since undergone a socio-economic shift away from fishing-dependent livelihoods to tourism (Diedrich, 2007), Sarteneja is a notable exception. Tourism in Sarteneja is minimal compared with other coastal communities and more than 80% of households are still dependent on fishing (SACD, 2009). However, there is strong interest in developing the tourism industry in the community, particularly given the serious decline in both lobster and conch fisheries, the two most important commercial species (SACD, 2009). In order to support both marine conservation and the community, BV undertakes a number of activities in Sarteneja, including environmental education in schools, sustainable livelihood activities tied to invasive lionfish (developing markets for and training fishers to handle lionfish; engaging women in lionfish jewellery production), and providing support for the Sarteneja Homestay Group. Since 2010, the Sarteneja Homestay Group has earned more than $250,000 USD in income from the BV volunteer tourism programme (BV, 2015), while other individuals in the community have benefited directly through employment (e.g. as cooks, boat drivers) and indirectly through spin-off economic benefits from volunteers (e.g. in shops, restaurants). The majority of local residents believe BV has increased economic opportunities in the community, although these opportunities are perceived as unevenly distributed (Ravensbergen, 2016).

This paper is based on data collected in 2013 and 2015, including 48 semi-structured interviews and four focus groups. This consisted of 38 semi-structured interviews and 1 focus group with volunteers in 2013 and 10 semi-structured interviews and 3 focus groups (1 with fishers, 1 with members of the Sarteneja Homestay Group, and 1 with high school students) in 2015. Two of the authors also conducted participant observation, including during one volunteer expedition in 2013 and for a total of six months in the village of Sarteneja (three months in 2013, three months in 2015). Interviewees included: volunteers (10), BV staff (8), members of the Sarteneja Homestay Group (9), representatives of two government agencies (3) and other local non-government organisations (NGOs) (4), and community members (14). Interviews and focus groups focused on a range of topics; those relevant for this paper include: perceptions of BV data collection activities, perceptions of BV conservation activities in the community, and the volunteer experience. Interviews were digitally recorded, translated from Spanish if necessary, transcribed in full, and then analysed for key themes using NVivo (qualitative data analysis software) (see Cope 2005). Interview excerpts are presented in order to illustrate key themes (i.e. widely shared perspectives discussed by multiple respondents), with respondents referred to by code (e.g. VT 01) to protect confidentiality.

Results

Volunteer skill-building and giving back through science

Two of the more common motivations expressed by conservation volunteers are the desire to support conservation efforts ('do good') and to gain practical skills and experience (Brondo, 2015; Campbell & Smith, 2005; Galley & Clifton, 2004). BV volunteers are no exception; the desire to gain practical experience that will augment their formal education and help to increase the competitiveness of their CVs is commonly reported.

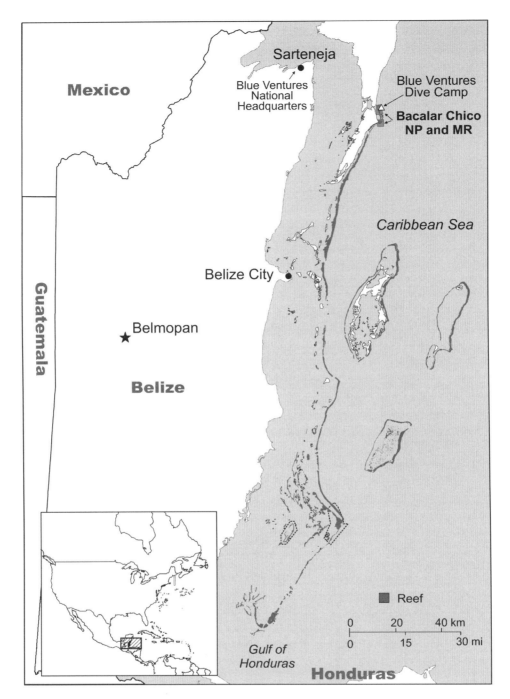

Figure 1. Map of study area in Belize.

... the primary reason for me doing anything like this was to bulk up my CV ... it's all well and good having a degree but if you don't have any practical experience putting these practices into use then it's no use having your degree. (VT 04)

Like volunteers in many other projects, this CV-building motivation is often coupled with an expressed desire to contribute, specifically to conservation and environmental management through data collection. Several volunteers commented on their contribution in this way. For example:

I wanted to get away but I didn't want it to be just a holiday, I wanted to be doing something [useful]. The fact that ... the research you do is actually going towards something, it's not just being discarded. (VT 07)

Unlike some other volunteer conservation projects, in which volunteers see their contribution as providing economic support to incentivise local communities to support conservation (e.g. Gray & Campbell, 2007), these volunteers conceptualise their data collection efforts specifically as a means of 'doing good'.

Furthermore, there is value attached to the scientific rigour and validity of the volunteer-collected data. Several volunteers noted BV's science specifically (rather than their conservation efforts more generally) as significant. 'I definitely like the scientific rigour ... and that their numbers are being used for something' (VT 03). Another volunteer noted that he chose the BV programme because 'these guys [BV] had published papers based on the work that they'd done, that to me as a scientist, peer reviewed papers are rock solid' (VT 08). BV staff similarly emphasise the quality of the data they collect. 'What I really like about BV is that the science really is the main goal ... our research is pretty rigorous, our training is very rigorous and definitely what I haven't found in any other [volunteer] organization before' (BV 06). The Fisheries Department was also in agreement on this point. 'We've had organizations like that before [whose data was not useful] ... but we have looked at the data that BV has been giving us and it's very good ... the data generated [by BV] will be helpful' (G 38).

In terms of global citizenship, the emphasis on doing science to 'do good' reflects a fairly soft, neoliberal mode of citizenship. Volunteers' efforts to be caring through their scientific contributions are simultaneously efforts to build their own skills and consume a scientifically legitimate experience (Cousins et al., 2009). However, the science produced by BV volunteers is perceived as more than just a tourist attraction or science for its own sake; the government agency responsible for managing the BCMRNP perceives the data as 'helpful'. This suggests that, consistent with lessons from citizen science, the BV volunteer science programme is supporting conservation policy and management.

Who can produce knowledge for conservation?

Not only do volunteers perceive their contribution as scientific knowledge production, they are also perceived as capable knowledge producers by a range of actors in Belize, including representatives of multiple government agencies, community groups, BV staff, and local residents. In some cases, local actors perceive foreign volunteers as better qualified than local residents to participate in scientific data collection. One government representative said, 'Sometimes it might be that foreigners are better able to do that than locals ... ' (G 33). Similarly, during a focus group a member of the Sarteneja Homestay Group noted, 'The volunteers are prepared, we don't have qualified students here [in Sarteneja] to do these things [data collection in BCMRNP].' In contrast, several local residents described their own practical, experience-based knowledge, suggesting this was often overlooked (in all conservation activities, not just BV volunteer tourism). For example, one representative of a local NGO said, 'the experts are the experts and you can't say anything' (NG 22), while a villager explained that 'what needs to be done is to get more community members involved ... [rather than] just by hearing it from the mouth [of scientists/conservationists]' (LR 44). This perception of a division between science and local knowledge is common, including in other areas of Belize (Gray, 2016). These results suggest that volunteers are perceived as donating expertise rather than learning with local residents, consistent with soft rather than critical global citizenship. They also suggest that local residents are not involved as participants in scientific research, contrary to lessons from citizen science.

Even though some local residents and government staff perceive foreign volunteers as better qualified than local residents to contribute to data collection activities, the nature of this qualification is not specified. One BV staff member indicated that a secondary school education is the minimum requirement.

> I think it would be difficult [to participate in data collection at BCDC] if they didn't have any type of education, but as long as you've got secondary school education ... you certainly don't need to be a marine biologist to be a volunteer. (BV 01)

Many respondents perceive volunteers as appropriate knowledge producers simply because they have the resources to do so, not because they are better qualified. For example, in discussing the data collection that volunteers undertake in the Bacalar Chico Marine Reserve, one fisher stated, ' ... it's good because fishers don't have time for this and they [volunteers] are doing things that wouldn't otherwise be done, like research' (focus group 4). As another villager noted, ' ... the volunteers are doing what we are not [able to do] in our own place' (LR 25). One respondent from a local community group characterised the role of volunteer knowledge production in this way:

> ... we are a small country and because as a government, as Belize, we don't have the resources to manage these protected areas, NGOs ... play that role in accessing funds that are from international funders and other sources that actually contribute to the sustainable management of these protected areas. (NG 26)

This sentiment is echoed by the Fisheries Department: 'It [volunteer data collection] is definitely beneficial ... the data is very expensive because you need man power, you need fuel and all those things and if the data quality submitted is good it's a plus for us' (G 38). One BV staff member explained: 'You either have the professionals doing it and you only have a little bit of data collected or you have 20 volunteers coming in and doing it and you have 20 times more data ... We need more rather than a tiny little set of solid professional data' (BV 20). Most volunteers are fully cognizant of their role in this respect. For example, one volunteer stated,

> ... in developing countries with not a big budget for things like environmental monitoring ... it kind of has to be tourists doing it and getting something out of doing it because otherwise it's not going to happen ... it's a way for people from very privileged rich backgrounds to give something back to poorer countries and that works on the community front as well as on the environmental front I think. (focus group 1)

Two aspects of 'giving back' are worth noting in relation to this case. First, the two components of 'giving back' noted by the volunteer – the 'community front' and the 'environmental front' – are geographically segregated in the case of BV. 'Giving back' to the community includes volunteering activities in Sarteneja (e.g. teaching classes) and providing economic benefits for the community. 'Giving back' through knowledge production includes data collection and monitoring activities undertaken at BCMRNP. As illustrated in Figure 1, BCMRNP is removed from the village – while a few fishers from Sarteneja occasionally visit the marine reserve, other villagers do not. Local residents do not participate in BV data collection activities, they do not see the volunteers collect data, and most of them have no direct experience in the place where data collection occurs. As one villager explained, ' ... never have I gone there [BCMRNP] ... because to go there is expensive' (LR 24). A BV staff member noted this challenge: 'Bacalar Chico is fished by [only] a handful of Sarteneja fishers, so I would say that our community work is not very closely linked with our work in Bacalar Chico ... ' (BV 53).

Second, giving back includes knowledge production through environmental monitoring – that what constitutes a 'global citizen' is in part a person who participates in knowledge production for the 'global good' – in this case, environmental conservation. However, once again this represents a soft conception of global citizenship. Volunteers embrace rather than question an unequal world system that presents them with the opportunity to 'give back' simply because they have the resources (time, money) to do so. Moreover, other actors (NGO staff, government representatives) also strategically embrace this model. The Fisheries Department does not perceive volunteer tourists as doing something they are incapable of doing, but rather something they lack the resources to do – volunteers are indirectly helping the state, through an NGO-managed project. This neoliberal rationale for volunteer tourism as a means to support environmental monitoring is identical to that for citizen science: volunteers can provide resources that scientists, government managers, and NGOs lack.

While foreign volunteers may be better educated than many residents of Sarteneja, it is access to resources (time, money) rather than skill or expertise that enables volunteers to participate in environmental monitoring in BCMRNP. In fact, the BV programme is designed to teach the necessary skills (fish identification, scuba diving) and does not require volunteers to have these beforehand.

> When I first started working here, I felt a bit sceptical about the way we took people in who weren't even divers yet ... surely if we just took only advanced divers we could get a lot more science done, but ... you can charge them more if you are training them to dive. You got to make money to run the projects so it has to be like that. (BV 01)

Some people involved with the BV programme reflected on the potential to connect the two spheres of 'giving back' – to direct some funding toward scholarships for local students to participate in expeditions alongside volunteers, for example.

> What can we do? We [could] give scholarships ... one person every expedition, we can take them from nothing to dive masters giving them professional qualifications, we can give them survey techniques ... and give them the skills that they need, not only that but the experience with working with international scientists. (BV 21)

However, this is as yet an unrealised possibility, to the frustration of this staff member. In his view, given the emphasis BV places on community benefits, the involvement of local residents in the scientific work should be an essential component. As another BV staff member observed, ' ... the main purpose of our presence in Sarteneja is community involvement, so capacity building ... the scientific part is the most important one' (BV 06). Several people (including volunteers, BV staff, and local residents) observed that BV does not have local residents working in administration or scientific roles. 'One of the huge problems in Belize is that many of the organizations are not managed by local people or the project managers aren't local or the research officer or the science officers aren't local' (BV 53). Within this programme and community, scientific research and monitoring is currently viewed as an activity undertaken by foreign individuals (whether volunteers or BV staff).

The opportunities identified by BV staff (to develop a scholarship and better integrate local residents into the scientific aspects of the programme) could help to challenge the conception of soft global citizenship implicit in actors' understanding of this volunteer project, by allowing opportunities for volunteers and local residents to learn with one another. This would also build on insights from citizen science, which suggest that all citizens should have the opportunity to participate and that outcomes for social-ecological systems are likely to be better when local residents feel engaged in the science associated with resource management processes.

Local residents as participants or recipients?

While there are perhaps practical solutions for reducing the 'resources divide', such as the provision of scholarships to support local participation, this would not address a more fundamental challenge of how volunteers and local residents are conceptualised as programme participants and recipients, respectively. Many volunteers perceive themselves as helping to teach local residents, and local residents as needing to learn. '[Blue Ventures is] working with the community to educate them so that they understand why [they can't fish], so it's for their own good' (VT 11). Another volunteer said:

> [My homestay mum] was saying that a lot of the fishermen ... they kind of resent the fact that this outsider group is coming and telling them how to do things. The reason why it's us telling them is because we have the education to be able to do these surveys and the money to do these surveys, which they don't and we can give them that information and make decisions based on that information which they don't have but I can see how it seems like such a kind of colonialist view. (VT 08)

Whether or not they acknowledged the problematic, colonial perspective, many volunteers discussed themselves as knowledge generators and providers, and local residents as recipients of this knowledge. In addition to including residents in scientific monitoring, another possibility for challenging this conceptual distinction would be to employ Belizeans as experts. As one BV staff member explained:

> We would like to employ more local people ... I think boat captains is the skilled staff that we have but we could have a field scientist or project manager or community officer or development officer, all of those could be filled by Belizeans and hopefully will be in the next 5 years it will be more Belizean and I won't have a job. (BV 53)

There is thus both interest and potential within BV to challenge this divide between volunteers as knowledge producers and local residents as knowledge recipients. As noted above, many local residents perceive themselves as experts – as knowledge producers and knowledge holders – just not as *scientific* experts. However, the production of knowledge through volunteer tourism – specifically, participation in environmental monitoring and scientific data collection in BCMRNP – currently serves to separate foreign volunteers from local residents.[1]

In order to foster critical global citizenship, the BV programme will need to move away from a model in which NGOs and foreign volunteers teach local residents to an approach in which all actors learn with and from one another. Recognising the willingness and capacity of local residents to participate in producing 'universal' scientific knowledge is one approach, while recognising, valuing, and integrating local expertise (e.g. by having local residents teach volunteers and/or contribute to design of scientific research projects) would be another. Citizen science research suggests that providing opportunities for local residents (and not just volunteers) to contribute to, collaborate on, or even co-create scientific research could enhance outcomes for all individuals involved (e.g. by increasing skills), as well as for science (by building on local knowledge) and social-ecological systems (e.g. by generating trust among stakeholders and better supporting management strategies) (Shirk et al., 2012).

Discussion and conclusion

Conservation volunteering is a particular kind of volunteer tourism in which 'doing good' is not just about caring for other people and places but also about generating scientific knowledge to support conservation efforts. If global citizenship entails a moral obligation to 'do good', then it is important to consider whether and how the enactment of this moral obligation intersects with ways of knowing. As the case of the BV volunteer programme demonstrates, producing scientific knowledge is central to how participants understand their contribution. However, the current configuration of the BV programme separates those who generate knowledge for conservation (BV staff and volunteers) from those who are the recipients of this knowledge (local residents, as well as the state). One danger of volunteer tourism is the inherent potential to reinforce rather than challenge unequal relationships, by involving the 'better off' in helping the 'worse off' (Lyons et al., 2012; Sin, 2009). Similarly, as this case demonstrates, it is important to consider whether volunteer tourism creates a situation in which those who are 'more knowledgeable' share expertise or provide information for those who are 'less knowledgeable'.

This divide between the 'more' and 'less' knowledgeable can be traced in part to tensions between science and local knowledge as legitimate ways of knowing – tensions that are widespread in Belize (and elsewhere), as the Belizean state prioritises science-based decision-making for marine management (Gray, 2016). However, the knowledge divide is also partly a function of the 'resources divide' underlying volunteer tourism. Volunteers are not necessarily more knowledgeable than local residents; indeed, it is because they want to *acquire* scientific skills and experience, rather than because of the knowledge they already have, that volunteers participate in conservation tourism programmes like BV. It is the 'resources divide' (between the 'better off' and 'worse off') rather than differences in knowledge that positions volunteers as knowledge producers and local residents as knowledge recipients.

The neoliberal logic underlying volunteer tourism, which simultaneously positions volunteers as neoliberal subjects seeking to enhance their (scientific) skills (Lyons et al., 2012; Sin et al., 2015) and sells scientific experiences as a market-based solution for conservation problems (Brondo, 2015; Cousins et al., 2009), creates a situation in which the ability to pay is entwined with the capacity to know. By highlighting 'the coming together of consumption and citizenship' through science, Jasanoff (2004, p. 91) connects this ability to pay and capacity to know with citizenship. We argue that global citizenship and knowledge are co-produced through volunteer tourism; in this case, the ability to pay to participate in science differentiates global citizens (those who generate knowledge to help solve environmental problems) from local residents (those who receive knowledge about these problems). Rather than creating solidarity between volunteers and local residents, the volunteer experience leads volunteers to conclude that they are more knowledgeable about marine ecosystems and conservation than are local residents, a pattern consistent with other conservation volunteer programmes (Brondo, 2015; Gray & Campbell, 2007; Grimm, 2013). Moreover, this case is illustrative of the ways in which identities (especially related to race) are performed and constituted through knowledge-making activities associated with conservation (Sundberg, 2004), a point that warrants additional attention in the volunteer tourism literature. In some important ways, this case demonstrates the limitations of volunteer tourism as a means of developing critical global citizenship through scientific participation.

However, the co-production of knowledge and global citizenship through volunteer tourism need not reinforce unequal relations. Some volunteers, and most BV staff, indicated an awareness of this problematic outcome and a desire for change. Whether acknowledging a colonial mindset or offering specific suggestions and strategies for engaging Belizeans as equal participants in scientific practices, many respondents identified opportunities for developing more critical, inclusive models of global citizenship in which *all* participants in volunteer tourism can contribute to knowledge production and develop scientific skills. Respondents' suggestions echo those offered by Brondo (2015), who identifies three ways in which conservation volunteer tourism can move forward. First, through collaborative programme design, volunteers, NGOs, and local residents can engage in dialogue about conservation practice, knowledge, and values. The literature on citizen science also offers important lessons for conservation volunteering on this point, illustrating the different models for engaging citizens in science (from contributors to collaborators to co-creators) and identifying the potential positive outcomes for conservation decision-making and practice (Shirk et al., 2012). If volunteer tourism is supposed to support conservation, then there is much to be gained from engaging local residents as citizen scientists (Larson, Conway, Hernandez, & Carroll, 2016). This is not a simple or straightforward task, nor does citizen science offer all of the answers. Indeed, citizen science has also been critiqued for failing to sufficiently challenge knowledge hierarchies and unequal power relations, thereby preventing socially just forms of environmental knowledge production (Burke & Heynen, 2014). However, given the similarities between the fields, lessons emerging from studies of citizen science are instructive for volunteer tourism.

Second, Brondo encourages volunteer programmes to support a transformative learning experience (Coghlan & Gooch, 2011) and not just resume building. As Lyons et al. (2012, p. 370) note, 'while the traditional view of cosmopolitanism requires one to become conversant with other cultures while maintaining a level of reflexivity about one's own culture, neoliberal interpretations do not require such reflexivity'. Here we would add that conservation volunteer programmes should not just support the development of volunteers' scientific skills; they should also encourage learning about other ways of knowing (e.g. fishers' local knowledge) and reflexivity about who produces science and for what purpose.

Third, Brondo (2015) notes the marginalisation of host community residents in volunteer tourism experiences and advocates for their integration into volunteer activities. She suggests, as did several BV respondents, the creation of a stipend to support such local involvement. We agree that it is important to integrate local perspectives into conservation efforts, including in relation to science and knowledge production for conservation. If global citizenship is to extend to *all* individuals involved in volunteer tourism, then opportunities must be created for local residents and volunteers to learn with one another.

Finally, it is important to note that the majority of residents of Sarteneja appreciate the social and economic benefits provided by volunteer tourism and value the opportunity to learn from volunteers through cross-cultural exchange (Ravensbergen, 2016). By drawing attention to the co-production of knowledge and global citizenship, we wish to highlight both opportunities and limitations of volunteer conservation programmes. If volunteer tourism is to produce critical global citizens through programmes that 'move away from exclusivity of knowledge' (Brosnan et al., 2015, p. 97), then both scholars and practitioners will need to continue to engage critically and constructively in moving the field forward.

Note

1. After the research for this paper was completed, Blue Ventures hired a Belizean Research Assistant. He holds a BS in Marine Biology and part of his job entails analysing data collected in BCMRNP.

Acknowledgements

We thank Louis Pena for his invaluable support as a research assistant, the staff of Blue Ventures, especially Jennifer Chapman and Marc Fruitema, and all research participants for sharing their time and perspectives. We also thank two anonymous reviewers for their helpful comments.

Funding

This work was supported by Social Sciences and Humanities Research Council of Canada [Grant Number 430-2013-000310].

Disclosure statement

No potential conflict of interest was reported by the authors.

References

Alexander, Z. (2012). International volunteer tourism experience in South Africa: An investigation into the impact on the tourist. *Journal of Hospitality Marketing and Management*, *21*(7), 779–799.

Andreotti, V. (2006). Soft versus critical global citizenship education. *Policy & Practice – A Development Education Review*, *3*, 40–51. doi:10.1057/9781137324665_2

Barbieri, C., Santos, C. A., & Katsube, Y. (2012). Volunteer tourism: On-the-ground observations from Rwanda. *Tourism Management*, *33*(3), 509–516. doi:10.1016/j.tourman.2011.05.009

Bianchi, R. V., & Stephenson, M. L. (2014). *Tourism and citizenship: Rights, freedoms and responsibilities in the global order*. Abingdon: Routledge.

Blue Ventures. (2015). *Conservation tourism: Driving conservation through sustainable tourism*. Retrieved February 6, 2017, from https://blueventures.org/impact/factsheets

Blue Ventures. (2016). *About Blue Ventures*. Retrieved November 1, 2016, from https://blueventures.org/about

Blue Ventures. (2017). *Who volunteers with Blue Ventures?* Retrieved February 6, 2017, from https://blueventures.org/volunteer/belize/#volunteer-profile

Bonney, R., Shirk, J. L., Phillips, T. B., Wiggins, A., Ballard, H. L., Miller-Rushing, A. J., & Parrish, J. K. (2014). Next steps for citizen science. *Science*, *343*(6178), 1436–1437. doi:10.1126/science.1251554

Borland, K., & Adams, A. E. (2013). Introduction. In K. Borland, & A. E. Adams (Eds.), *International volunteer tourism: Critical reflections on good works in Central America* (pp. 1–5). New York, NY: Palgrave MacMillan.

Brightsmith, D. J., Stronza, A., & Holle, K. (2008). Ecotourism, conservation biology, and volunteer tourism: A mutually beneficial triumvirate. *Biological Conservation*, *141*(11), 2832–2842. doi:10.1016/j.biocon.2008.08.020

Brondo, K. V. (2015). The spectacle of saving: Conservation voluntourism and the new neoliberal economy on Utila, Honduras. *Journal of Sustainable Tourism*, *23*(10), 1405–1425. doi:10.1080/09669582.2015.1047377

Brosnan, T., Filep, S., & Rock, J. (2015). Exploring synergies: Hopeful tourism and citizen science. *Annals of Tourism Research*, *53*(0), 96–98. doi:10.1016/j.annals.2015.05.002

Burke, B. J., & Heynen, N. (2014). Transforming participatory science into socioecological praxis: Valuing marginalized environmental knowledges in the face of the neoliberalization of nature and science. *Environment and Society: Advances in Research*, *5*, 7–27. doi:10.3167/ares.2014.050102

Butcher, J., & Smith, P. (2010). 'Making a difference': Volunteer tourism and development. *Tourism Recreation Research, 35*(1), 27–36.

Butcher, J., & Smith, P. (2015). *Volunteer tourism: The lifestyle politics of international development*. London: Routledge.

Cameron, J. D. (2013). Grounding experiential learning in 'thick' conceptions of global citizenship. In R. Tiessen & R. Huish (Eds.), *Globetrotting or global citizenship: Perils and potential of international experiential learning* (pp. 21–42). Toronto, ON: University of Toronto Press.

Campbell, L. M., & Smith, C. (2005). Volunteering for sea turtles? Characteristics and motives of volunteers working with the Caribbean Conservation Corporation in Tortuguero, Costa Rica. *MAST, 3/4*(2/1), 169–194.

Chapman, J. K. (2014). *Status of marine resources in Bacalar Chico Marine Reserve 2013 Blue Ventures conservation report*. London: Blue Ventures.

Cho, L. (2005). Marine protected areas: A tool for integrated coastal management in Belize. *Ocean & Coastal Management, 48*, 932–947.

Clifton, J., & Benson, A. (2006). Planning for sustainable ecotourism: The case for research ecotourism in developing country destinations. *Journal of Sustainable Tourism, 14*(3), 238–254.

Coghlan, A., & Gooch, M. (2011). Applying a transformative learning framework to volunteer tourism. *Journal of Sustainable Tourism, 19*(6), 713–728.

Conran, M. (2011). They really love me!: Intimacy in volunteer tourism. *Annals of Tourism Research, 38*(4), 1454–1473. doi:10.1016/j.annals.2011.03.014

Cope, M. (2005). Coding qualitative data. In I. Hay (Ed.), *Qualitative research methods in human geography* (pp. 223–233). Oxford: Oxford University Press.

Cousins, J. A. (2007). The role of UK-based conservation tourism operators. *Tourism Management, 28*(4), 1020–1030. doi:10.1016/j.tourman.2006.08.011

Cousins, J. A., Evans, J., & Sadler, J. (2009). Selling conservation? Scientific legitimacy and the commodification of conservation tourism. *Ecology and Society, 14*(1), 32. Retrieved from http://www.ecologyandsociety.org/vol14/iss1/art32

Daly, K. (2013). Who is a global citizen? Manifestations of theory in practice. In K. Borland & A. E. Adams (Eds.), *International volunteer tourism: Critical reflections on good works in Central America* (pp. 69–78). New York, NY: Palgrave MacMillan.

Dickinson, J. L., Shirk, J., Bonter, D., Bonney, R., Crain, R. L., Martin, J., … Purcell, K. (2012). The current state of citizen science as a tool for ecological research and public engagement. *Frontiers in Ecology and the Environment, 10*(6), 291–297.

Diedrich, A. (2007). The impacts of tourism on coral reef conservation awareness and support in coastal communities in Belize. *Coral Reefs, 26*(4), 985–996.

Donnelly, A., Crowe, O., Regan, E., Begley, S., & Caffarra, A. (2014). The role of citizen science in monitoring biodiversity in Ireland. *International Journal of Biometeorology, 58*(6), 1237–1249.

Edwards, R. (2014). The 'citizens' in citizen science projects: Educational and conceptual issues. *International Journal of Science Education, Part B, 4*(4), 376–391.

Follett, R., & Strezov, V. (2015). An analysis of citizen science based research: Usage and publication patterns. *Plos One, 10*(11), e0143687. doi:10.1371/journal.pone.0143687

Galley, G., & Clifton, J. (2004). The motivational and demographic characteristics of research ecotourists: Operation Wallacea volunteers in Southeast Sulawesi, Indonesia. *Journal of Ecotourism, 3*(1), 69–82.

Government of Belize. (2015). *National protected areas system plan, Revised Edition*. Retrieved February 6, 2017, from http://protectedareas.gov.bz/technical-documents

Gray, N. J. (2016). The role of boundary organizations in co-management: Examining the politics of knowledge integration in a marine protected area in Belize. *International Journal of the Commons, 10*(2), 1013–1034. doi:10.18352/ijc.643

Gray, N. J., & Campbell, L. M. (2007). A decommodified experience? Exploring aesthetic, economic, and ethical values for volunteer ecotourism in Costa Rica. *Journal of Sustainable Tourism, 15*(5), 463–482.

Grimm, K. E. (2013). Doing 'conservation': Effects of different interpretations at an Ecuadorian volunteer tourism project. *Conservation and Society, 11*(3), 264–276.

Irwin, A. (1995). *Citizen science*. London: Routledge.

Jansujwicz, J. S., Calhoun, A. J. K., & Lilieholm, R. J. (2013). The Maine Vernal Pool mapping and assessment program: Engaging municipal officials and private landowners in community-based citizen science. *Environmental Management, 52*(6), 1369–1385. doi:10.1007/s00267-013-0168-8

Jasanoff, S. (2004). Science and citizenship: A new synergy. *Science and Public Policy, 31*(2), 90–94.

Johnson, M. F., Hannah, C., Acton, L., Popovici, R., Karanth, K. K., & Weinthal, E. (2014). Network environmentalism: Citizen scientists as agents for environmental advocacy. *Global Environmental Change, 29*, 235–245. doi:10.1016/j.gloenvcha.2014.10.006

Larson, L., Conway, A., Hernandez, S., & Carroll, J. (2016). Human-wildlife conflict, conservation attitudes, and a potential role for citizen science in Sierra Leone, Africa. *Conservation and Society, 14*(3), 205–217. doi:10.4103/0972-4923.191159

Lave, R. (2012). Neoliberalism and the production of environmental knowledge. *Environment and Society, 3*(1), 19–38. doi:10.3167/ares.2012.030103

Leach, M., & Scoones, I. (2005). Science and citizenship in a global context. In M. Leach, I. Scoones, & B. Wynne (Eds.), *Science and citizens: Globalization and the challenge of engagement* (pp. 15–38). London: Zed Books.

Leach, M., Scoones, I., & Wynne, B. (Eds.). (2005). *Science and citizens: Globalization and the challenge of engagement*. London: Zed Books.

Lorimer, J. (2009). International conservation volunteering from the UK: What does it contribute? *Oryx, 43*(3), 352–360. doi:10.1017/s0030605309990512

Lorimer, J. (2010). International conservation 'volunteering' and the geographies of global environmental citizenship. *Political Geography, 29*(6), 311–322. doi:10.1016/j.polgeo.2010.05.004

Lyons, K., Hanley, J., Wearing, S., & Neil, J. (2012). Gap year volunteer tourism: Myths of global citizenship? *Annals of Tourism Research, 39*(1), 361–378.

McGehee, N. G. (2012). Oppression, emancipation, and volunteer tourism: Research propositions. *Annals of Tourism Research, 39*(1), 84–107. doi:10.1016/j.annals.2011.05.001

Medina, L. K. (2015). Governing through the market: Neoliberal environmental government in Belize. *American Anthropologist, 117*(2), 272–284.

Miller-Rushing, A., Primack, R., & Bonney, R. (2012). The history of public participation in ecological research. *Frontiers in Ecology and the Environment*, *10*, 285–290. doi:10.1890/110278

Mostafanezhad, M. (2013). The geography of compassion in volunteer tourism. *Tourism Geographies*, *15*(2), 318–337.

Mostafanezhad, M. (2014). *Volunteer tourism: Popular humanitarianism in neoliberal times*. New York, NY: Routledge.

Pandya, R. E. (2012). A framework for engaging diverse communities in citizen science in the US. *Frontiers in Ecology and the Environment*, *10*(5), 314–317.

Ravensbergen, S. (2016). *Marine conservation and volunteer tourism: Examining community perceptions in Sarteneja, Belize (MA Thesis)*. University of Guelph.

Raymond, E. M., & Hall, C. M. (2008). The development of cross-cultural (Mis)Understanding through volunteer tourism. *Journal of Sustainable Tourism*, *16*(5), 530–543. doi:10.1080/09669580802159610

Rogerson, J. M., & Slater, D. (2014). Urban volunteer tourism: Orphanages in Johannesburg. *Urban Forum*, *25*(4), 483–499. doi:10.1007/s12132-014-9240-6

Sarteneja Alliance for Conservation and Development (SACD). (2009). *Sarteneja tourism development plan*. Sarteneja: Author.

Shirk, J. L., Ballard, H. L., Wilderman, C. C., Phillips, T., Wiggins, A., Jordan, R., … Bonney, R. (2012). Public participation in scientific research: A framework for deliberate design. *Ecology and Society*, *17*(2). doi:10.5751/ES-04705-170229

Simpson, K. (2005). Dropping out or signing Up? The professionalisation of youth travel. *Antipode*, *37*(3), 447–469.

Sin, H. L. (2009). Volunteer tourism – 'Involve me and I will learn'? *Annals of Tourism Research*, *36*(3), 480–501. doi:10.1016/j.annals.2009.03.001

Sin, H. L. (2010). Who are we responsible to? Locals' tales of volunteer tourism. *Geoforum*, *41*(6), 983–992. doi:10.1016/j.geoforum.2010.08.007

Sin, H. L., Oakes, T., & Mostafanezhad, M. (2015). Traveling for a cause: Critical examinations of volunteer tourism and social justice. *Tourist Studies*, *15*(2), 119–131. doi:10.1177/1468797614563380

Stainton, H. (2016). A segmented volunteer tourism industry. *Annals of Tourism Research*, *61*, 256–258. doi:10.1016/j.annals.2016.09.011

Statistical Institute of Belize. (2013). *Abstract of statistics, Belize*. Retrieved May 2, 2016, from http://www.sib.org.bz/Portals/0/docs/publications/abstract/AbstractofStatistics_2013.pdf

Stevens, M., Vitos, M., Altenbuchner, J., Conquest, G., Lewis, J., & Haklay, M. (2014). Taking participatory citizen science to extremes. *IEEE Pervasive Computing*, *13*(2), 20–29.

Sundberg, J. (2004). Identities in the making: Conservation, gender and race in the Maya Biosphere Reserve, Guatemala. *Gender, Place and Culture*, *11*(1), 43–66.

Vrasti, W., & Montsion, J. M. (2014). No good deed goes unrewarded: The values/virtues of transnational volunteerism in neoliberal capital. *Global Society*, *28*(3), 336–355. doi:10.1080/13600826.2014.900738

Wearing, S., & McGehee, N. G. (2013). Volunteer tourism: A review. *Tourism Management*, *38*, 120–130. doi:10.1016/j.tourman.2013.03.002

Weaver, D. (2015). Volunteer tourism and beyond: Motivations and barriers to participation in protected area enhancement. *Journal of Sustainable Tourism*, *23*(5), 683–705.

World Travel & Tourism Council (WTTC). (2016). *Travel & tourism economic impact 2016 Belize*. London: WTTC.

Yamamoto, D., & Engelsted, A. K. (2014). World wide opportunities on organic farms (WWOOF) in the United States: Locations and motivations of volunteer tourism host farms. *Journal of Sustainable Tourism*, *22*(6), 964–982. doi:10.1080/09669582.2014.894519

Zeddies, M., & Millei, Z. (2015). 'It takes a global village': Troubling discourses of global citizenship in United Planet's voluntourism. *Global Studies of Childhood*, *5*(1), 100–111.

Volunteer tourism in Romania as/for global citizenship

Cori Jakubiak and Iulia Iordache-Bryant

ABSTRACT
This paper examines short-term, volunteer tourism in Romania as a form of education for global citizenship. Drawing upon data collected through participant observation and qualitative interviews with numerous stakeholders, we argue that global citizenship education in this programme not only aligns with neoliberal governmentality, as others have deftly pointed out vis-à-vis volunteer tourism more broadly, but also has troubling implications for volunteering as civic engagement. Due to a short length of stay, a significant language barrier, limited professional or technical expertise, and a lack of knowledge of the service context, what volunteer tourists were able to do in a Romanian intentional community research context, was extremely limited: they primarily played with children. And while this childcare function occasionally offered pragmatic assistance to local community members, volunteers' evolving understandings of the purposes and role of volunteering in global citizenship contrasted sharply with other understandings of volunteering – particularly those that frame volunteering as a component of democratic citizenship education.

Introduction

Volunteer tourism is defined as short-term, alternative travel that combines voluntary service with holidaying (Wearing, 2001). Associated with global civic action (cf. McBride & Sherraden, 2007), volunteer tourism has been incorporated into postsecondary education through course credit reciprocity (e.g. Global Service Corps, 2008), and US federal legislation deems the practice a Global South development initiative (H.R. 1388 - Serve America Act, 2009). Many college-bound students now participate in volunteer tourism to increase the competitive advantage of their applications (Heath, 2007), and some companies offer sabbatical leave opportunities for employees to be volunteer tourists (Jakubiak, 2015).

One form of value linked to volunteer tourism is the notion of global citizenship (Butcher & Smith, 2015). Although sponsoring non-governmental organisations (NGOs) and corporations generally do not operationalise the term, global citizenship, it and similar monikers appear frequently in promotional literature. In illustration, WorldTeach (n.d.), tells prospective volunteers: 'The world needs wider access to education and more compassionate global citizens'. Similarly, United Planet (2016) advertises: 'Our [volunteer tourism] programs emphasize authenticity: Get outside your bubble, jump into a real experience, become a global citizen'. These discourses suggest that participating in volunteer tourism automatically renders one a global citizen. Yet what, exactly, constitutes global citizenship is rarely addressed in these same discourses.

This paper examines volunteer tourism as a form of global citizenship education. Drawing upon data collected through participant observation in a US-based, NGO-sponsored volunteer tourism programme in Romania (marketed as an orphanage), we trace the contours of global citizenship by addressing the following research questions: (1) What are some aspects of global citizenship if participation in this or similar programmes is adequate preparation for it? (2) What is the purpose of volunteering in global citizenship education? (3) What is the relation (if any) between global citizenship education and other kinds of citizenship education?

Our findings suggest that global citizenship as operationalised through a volunteer tourism programme in Romania not only aligns with neoliberal governmentality, as others have noted vis-à-vis volunteer tourism more broadly (cf. Vrasti, 2013), but also has troubling implications for *volunteering* as civic engagement. Due to a short length of stay, a significant language barrier, limited professional expertise, and a lack of knowledge of the service context, what volunteer tourists were able to do in Romania was limited: they primarily played with children. And while this childcare function offered some assistance to resident caregivers, it did not address the community's material needs or teach volunteers about Romania. Global citizenship education, in this context, did not meet the goals of even conservative democratic citizenship education programmes – those that stress volunteerism as a way to address material problems (Westheimer & Kahne, 2004). A feature of global citizenship, then, may be a diluted notion of what *volunteering* can do and who should be its primary beneficiaries.

For the purposes of this paper, we situate volunteer tourism theoretically within scholarship on neoliberal governmentality. This framework helps to account for why volunteer tourism need not meet any material goals yet is interpreted as meaningful civic action.

Volunteer tourism as life politics and governmentality

Recent scholarship on volunteer tourism ties the practice to ethical consumption. Like fair trade or cause-related shopping, volunteer tourism is a 'pay-to-participate' activity that serves as evidence of civic engagement. Jim Butcher and Peter Smith (2010) describe this linking of civic action with purchasing as lifestyle, or *life politics*: the expression of political agency through the purchase of select goods or services.

Studies in governmentality (e.g. Dean, 1999/2010; Rose, 1999) provide insight on the rise of life politics. Consequently, this work helps to account for why certain forms of social action – such as volunteer tourism – have come to seem right or commonsensical. While lifestyle research attends to how advertising and products interpellate people as consumers, governmentality studies examine the ways in which 'the lives of modern individuals are shaped by specific purposes and tasks assigned to them by authorities of various kinds' (Binkley, 2007, p. 111). These authorities include governments, credentialed experts, and educational institutions, all of whom employ *technics* (Inda, 2005) to generate a particular version of the world. Technics turn select observations into legible truths; they thus help to define the realities and discourses within (and upon which) which people are expected to act.

Numerous technics hail the contemporary subject as a volunteer. From George H.W. Bush's (1988) 'Thousand Points of Light' speech, in which the then-presidential nominee likened volunteerism to 'a thousand points of light in a broad and peaceful sky' to former president Barack Obama's (2009) inaugural 'Call to Service', US citizens, like many Global Northerners, are continually advised to address broad-scale problems through volunteerism. Other technics such as compulsory community service mandates (see Schwartz, 2010), college service-learning programmes, and even airport billboards (cf. The Foundation for a Better Life, 2014) produce a social arena in which volunteering is not only the proper response to concerns ranging from illiteracy to cancer, but also a critical means towards self-actualisation (cf. King, 2006).

This promotion of volunteering for its experiential benefits rather than its remedying effects suggests changes in modes of state governance. In his 1978–1979 lectures, Michel Foucault observed that as the modern state shifted its priorities from providing public services to assisting the mandates of capital, its processes of exercising sovereignty over its subjects were also changing (Burchell, Gordon, & Miller, 1991). Under conditions of advanced liberalism (Rose, 1999) in which Keynesian market interventions and state welfare programmes, *inter alia*, are seen as inhibiting free markets, citizens, too, are liberated from the state in new ways. Subjects are encouraged to view themselves as self-regulating actors in accordance with market rationalities, and the state's role is to provide opportunity spaces – from school voucher programmes to private health care markets – in which individuals can exercise these new freedoms (cf. Dean, 1999/2010).

It is in this context that volunteer tourism may be read as a form of governmentality. Since the welfare state is recast as the 'enabling' state (Rose, 1999, p. 142), a sovereign that generates the conditions for personal growth and self-optimising rather than mitigating capitalism's social effects, volunteer tourism fills a particular niche. It offers to meet needs previously addressed by the state while providing an unmatched, dramatic experience for personal growth (cf. Vrasti, 2013).

The links that we have presented here among volunteer tourism, life politics, and governmentality have numerous implications. They affect not only how people understand political agency in the present, but also what *volunteering* itself now means. In a society in which '[i]ndividuals are discouraged from seeing life in terms of any collectivist obligation or shared purpose, and encouraged to undertake their lives as projects of heightened individuality, self-reliance, and opportunity maximization' (Binkley, 2007, p. 119), volunteering is frequently seen as something one does for oneself and one's personal benefit rather than to meet external goals or needs. This helps to explain why volunteer tourists need not deliver tangible services to a host community but earn cultural capital for their efforts nevertheless. Among these forms of cultural capital is global citizenship.

Global citizenship

In addition to playing an important role in fostering neoliberal governmentality, volunteer tourism can be explained as part of a broader, cultural preoccupation with all things global. Anthropologist Anna Lowenhaupt Tsing (2000) has observed a rising interest in what she terms *globalism:* a 'kind of scaling practice ... [that] valorizes global connections, linkage and circulation ... and associates them with progress' (cited in Doerr, 2012a, p. 4). Globalism has strong roots in the field of education (Standish cited in Butcher & Smith, 2015), where '[o]ne manifestation of globalism is the suggestion [for students] to proactively adapt to the "globalizing world", which is assumed to be a given' (Doerr, 2012a, p. 2). Course-embedded travel, international service-learning, study abroad, and volunteer tourism all exemplify globalist projects. They offer to help students develop awareness of international issues through immersion in distant, seemingly spatially bound, communities.

Like the demands of neoliberal governmentality, projects to create global citizens are changing the contours of political engagement. Loosely defined as those 'who understand the interconnectedness of the world, engage in issues of global significance, and relate to those from other cultures' (Deardorff cited in Doerr 2012b, p. 10), global citizens are urged to care about (and act upon) issues that transcend national borders. They do this not through customary political avenues, however (such as exercising the vote), but by purchasing politically inflected products, curating socially conscious lifestyles, and engaging in globalist projects like volunteer tourism. As Jim Butcher and Peter Smith (2015) put it, global citizenship involves 'a reworking of the concept of citizenship not only spatially from nation to globe, but also politically from nation and polity to non-governmental organizations and consumption' (p. 89).

This reworked notion of citizenship can be seen in volunteer tourism promotional literature. In illustration, Cross-Cultural Solutions (2014) characterises political action in the present as follows:

> More than ever, people around the world want change. Change in the inequities that polarize. Change in the corrupt systems that prevent self-determination. Change in the unjust repression of entire populations. *But the change we all wish to see won't be realized through big, sweeping acts – not by governments, or armies, or the UN.* Instead, lasting change will be achieved through small, personal acts of kindness and selflessness, and through the spreading of tolerance and understanding between people and cultures. (emphasis added)

Suggested here is that democratically elected legislative bodies, peacekeeping teams, and multilateral actors play little to no role in resource distribution, human rights, or geopolitical security. Instead, individual, entrepreneurial-style civic action is the most effective way to address social problems.

The concepts that we have outlined here – neoliberal governmentality and global citizenship – situate our study of Romanian volunteer tourism topically and theoretically. They help to explain why the participants in our study did very little, materially, while volunteering abroad but ascribed significance to their service (and were rewarded for it) nevertheless. We turn to data that extends this argument following an explication of our study's methods.

Methods

Data for this study were collected between April and July 2014. Having received a grant to study volunteer tourism qualitatively, we selected to participate a programme run by a well-regarded, US-based NGO – an NGO unaffiliated with our institution. As of this writing, the NGO has existed for 15 years and offers programmes in 35 countries; it sends hundreds of volunteers abroad annually, mainly from the US and Canada (NGO staff member, personal communication, April 30, 2014). The NGO has received positive media attention from CNN and NPR, as well as an endorsement from an Oscar-winning Hollywood star. The NGO's website asserts that it is 'unlocking the potential of every person as a global citizen and catalyst to create a more peaceful, cohesive, and sustainable world', rhetoric common to many volunteer tourism sponsors.

We chose the NGO's Romania programme because the second author, a White, female Romanian undergraduate student in her early twenties, hailed from the province in which the programme is based and had previously volunteered there as a high school student (as a Romanian national, not a volunteer tourist). Her fluency in Romanian – combined with her familiarity with the region – presented a unique opportunity for us to gain local peoples' perspectives on volunteer tourism, which remain underrepresented in the literature (see Gray & Campbell, 2007; McGehee & Anderbeck, 2009 for exceptions).

This programme also typified numerous aspects of volunteer tourism. It had no foreign language requirements; participants did not need to possess any specific skills or credentials; its cost was comparable to other US-based NGOs' programmes; and the promotional language surrounding the programme – stressing volunteer impact, flexible travel times, authentic cultural encounters, simplistic ways of life, and personal safety – reflected volunteer tourism's dominant discourses (cf. Keese, 2011; Simpson, 2004). Our study's findings, then, though partial by design (Charmaz, 2006), offer insight on volunteer tourism as a broader phenomenon.

Prior to departure, we participated in a 1-hour phone interview with a US-based NGO staff member and a 1-hour, online webinar with approximately 10 other people (via a conference call). The webinar was geared towards anyone volunteering through the NGO that summer, so the information presented was applicable to programmes in Africa, South America, and Europe. Issues addressed included, but were not limited to, being lesbian, gay, bisexual, transgender, or queer abroad, the importance of drinking clean/bottled water in-country, and how to contact NGO support staff. No specific information about Romania or our programme was discussed. Some of the webinar participants were parents of volunteers, not the volunteers themselves; their questions tended to focus on volunteer safety.

We also received a package that included handouts on 'How to be a Sustainable Traveler', which recommended buying carbon off-sets; 'American Values as Perceived Abroad', which offered recommendations on how to respond to generalised anti-American sentiment (in any country); and 'Cross-Cultural Adjustment Activities', which recommended keeping a journal. None of these handouts were specific to Romania or our host site. Finally, we received t-shirts imprinted with the NGO's mission statement, which we were instructed to wear on our travel date to Romania.

For 10 days in July, 2014, we engaged in participant observation in a central Romanian intentional community that we term Celebrate Life. Associated with non-traditional, often utopian, living arrangements such ecovillages and communes, intentional communities

> differ from the society that surrounds them, because they are intentional and because they are communal. Because they are intentional, people who live in them are not neighbors by happenstance; they have chosen to live together. Because they are communal, they share things that neighbors do not normally share, such as wealth, property, food, labor, and sometimes even spouses. (Kamau, 2002, p. 17)

Given its communal nature, Celebrate Life was atypical in the emergent capitalist economy of Romania. An Eastern Orthodox intentional community, Celebrate Life was established in the early 1990s by a conservative Romanian priest. At the time of our study, it housed approximately 200 full-time residents: mothers and their children; senior citizens; dependent children; caregivers; and cognitively disabled adults, many of whom had been reared in Ceaușescu's state institutions. Most residents identified as Eastern Orthodox and attended on-site services twice daily; these services were announced throughout the campus by a tune played on wood instruments. The campus included a church, agricultural buildings, shared homes, a partially-built vocational school, a volunteer house, and a communal dining hall in which residents and volunteers ate meals, albeit at separate tables. Approximately 200 additional people from the surrounding village came to Celebrate Life daily for meals, religious services, and/or employment.

Despite the diversity of residents, visitors, and activities at Celebrate Life, the NGO marketed its programme there as occurring solely in an orphanage. It was only because of the second author's background that we knew about Celebrate Life's status prior to our arrival. In Romania, Celebrate Life is well-known for its ideology and alternative structural organisation; it operates through philanthropic activity and solicits donations on television. Moreover, many mothers live on-site with their children, so is not primarily an orphanage. Language on the NGO's website, however, did not mention these resources. Instead, its programme description stated:

> Helping vulnerable children is at the heart of [the NGO's] focus. [The NGO] strives to give these children a voice and to take charge of their immediate needs, and provide training and opportunities to take charge of their lives to have a better chance at a better future. Volunteers can play with the children who are in great need of care and attention, teach English, teach arts and crafts, and more.

This discourse demonstrates how volunteer tourism normalises active, entrepreneurial citizen-subjectivity (cf. Vrasti, 2013). If children are vulnerable, what they need is personal empowerment, not structural or policy changes that could channel more material resources their way.

Upon our arrival at Celebrate Life, we received a tour of the campus, basic Romanian lessons (in which the second author was used as a translator), and directions on where to locate games, sports equipment, and arts and crafts materials. Despite the NGO's claim that we would 'take charge of [resident children's] immediate needs', we and our fellow volunteers received little guidance on how to be of use. Resident children ate, lived, and, for the most part, recreated in separate dwellings; volunteer tourists' interactions with children were ad hoc. Consequently, volunteers' activities were unscheduled; they centred on playing cards, board games, and doing arts and crafts with youth or cognitively disabled adults; walking the Celebrate Life campus; pushing youngsters on swings; watching resident teens perform dances and/or skits; baking in the volunteer house; and visiting the nearby village bar.

Because the second author is Romanian, she evoked curiosity and excitement among Celebrate Life residents. When she would greet them in Romanian, they often registered surprise on their faces and would thereafter seek her out. She read books (in Romanian) with resident children, and we were invited (unlike other volunteer tourists) into adults' private spaces for tea, to look at pictures, and to make food. During these times, Romanian was spoken exclusively.

In addition to recording our activities and reflections in field notes, we conducted 16 semi-structured, 45–60 minute, open-ended interviews with volunteers and community members. Thirteen of these interviews were audiotaped and transcribed for analysis. Three of the interviews were not recorded (due to some Romanians' discomfort with audio-recordings, a possible legacy of having lived under Ceaușescu, but notes of these interviews were taken. Formal interview participants included eight volunteers (seven from the US and one from Spain), all of whom had found the NGO's programme through the Internet and chose it because of its time flexibility and lack of language requirement. We also interviewed an on-site volunteer coordinator from the UK and seven Romanian residents: one senior citizen, four caregivers, one mother, and one volunteer coordinator. Interviews with Romanians were conducted in Romanian; interviews with volunteers and expatriate NGO staff were conducted in English. The second author also took field notes by recording informal conversations with Celebrate Life residents and employees, many of whom declined to be interviewed or expressed discomfort at the prospect.

Indeed, our attempts to learn about locals' perceptions of volunteer tourism (a call issued by many volunteer tourism scholars; see McGehee & Andereck, 2009) were rendered difficult by the sociohistorical context of conducting research in a post-dictatorship. During a recorded interview, for example, one caregiver underreported how long she had worked at Celebrate Life. As a result, we discarded that interview and developed concerns about whether employees felt coerced into study participation. We subsequently stopped requesting formal interviews from employees and relied on informal conversations to gain their perspectives.

While study participants' social locations reflected a variety of perspectives, related themes emerged in the data corpus. These themes became more trustworthy as we triangulated interview data with field notes, promotional literature, and member checks. Data were analysed using a constructivist grounded theory approach (cf. Charmaz, 2006). Following initial coding, writing analytical memos, and focused coding (Emerson, Fretz, & Shaw, 1995), all recursive, we identified three main themes that tied to our research questions. These were *bringing smiles*; *little listening to others' voices*; and *voluntary service as primarily about the self*, which we explicate in the following section.

Bringing smiles

Many researchers have presented evidence that volunteer tourism programmes equate bringing smiles, hugs, and joy to distant others with meaningful civic engagement (e.g. Guiney, 2017; Sinervo, 2011). This finding has been theorised as representing contemporary development formations, which stress well-being (Butcher & Smith, 2015) and personal intimacy (Conran, 2011) over economic or social transformation. Sam Binkley's (2007) work on happiness provides an additional, useful lens through which to view this phenomenon. Binkley writes that one manifestation of neoliberal governmentality is an 'intensifying discourse around human happiness, which is presented as … the rationally conceived end of a project of self-government outlined in a range of policy debates' (p. 121). National happiness indexes and the popularity of trade books such as *The Happiness Project* illustrate this trend. Similarly, volunteer tourists' understanding of bringing smiles to others as meaningful civic participation can be interpreted as reflecting new modes of neoliberal governance and the political obligations they entail.

Exemplifying this perspective was Gwen, a White, female, returning volunteer in her early twenties from the US Midwest. Gwen often took small groups of children from the Celebrate Life campus to a pool and out for pizza (at her own expense). She explained the purpose of her volunteering at Celebrate Life as follows:

> It's the reaction when I come back that I love, like, you know, when the kids see me and they get so happy and get so excited … The children just love, love to have volunteers come. Because they [the volunteers] plan activities for them; they do things. And just, you know, that [volunteers] planted a seed. And even though they might never come back and they might not see that seed grow, they still – it's still something there. And within that week that the volunteer came, they put smiles on [children's] faces. (July 14, 2014)

Shannon, a former Celebrate Life volunteer coordinator from the US Northeast and return volunteer, shared a similar perspective. Contrasting other kinds of volunteering with that done at Celebrate Life, Shannon said:

> Physical labor, you see – you see the end results. You put in the time and you see the end result, and you feel good because you're like, "Okay, see what I've done? Awesome." But then when you're working with children, which is one of the main things here, you may not be able to see the result besides a smile on the kid's face. (July 9, 2014)

While admitting that some volunteers become frustrated by their inability to see material service outcomes, Shannon, like Gwen, asserts that making children smile is an adequate outcome of volunteering.

Kristy, an on-site volunteer coordinator from the UK, spoke in related terms. In addition to bringing smiles, she explained, volunteer tourists provide hope for Celebrate Life's children through their personal examples. She said:

> I hope [volunteers] go away feeling hopeful about the people here, and hopeful that ... as individuals, we can make a difference to peoples' lives. Because, you know, you never know who's going to make that difference. We have had volunteers who have had a probably profound effect on some of the kids ... It gives the kids a bit of hope that, you know, if they work hard ... there is more out there than what is here ... We look at volunteers as opening the eyes of the kids. (July 12, 2014)

As these data excerpts suggest, volunteers' contributions to Celebrate Life are delivering smiles, hope, and imagined mobilities (Jakubiak, 2017) to children characterised as unhappy. This finding resonates with related research, which suggests that volunteer tourism prioritises *social inclusion* as a development aim. This emphasis places 'objective indices of wealth or unemployment alongside ... subjective notions such as self-esteem in its conception of human welfare' (Butcher & Smith, 2015, p. 46). Such a view equates raising peoples' self-images with bolstering their material security.

In another mode of accumulation or a different historical time period, volunteering as political action might once have be strictly defined as meeting others' material needs – for example, by serving food at a Black Panther school breakfast programme (Perlstein, 2008). Under the demands of neoliberal subjectivity, however, '[h]appiness is a form of rationality ... imposed on individuals through social policy initiatives aimed at the development and management of populations. To the extent that people are happy, it is believed, they are well governed and well administered' (Binkley, 2007, p. 113). The active, entrepreneurial citizen has a civic obligation to be happy. Efforts to make others happy, in turn, count as voluntary service in this frame.

Little listening to others' voices

A key component of justice-oriented, democratic citizenship education is building activist alliances. Although not sufficient on its own, volunteering can serve as a way to learn more about marginalised groups, their struggles, and the relations between privilege and subordination (Swalwell, 2013). A participant in this study, for example, shared how domestic volunteering had complicated his previous understanding of homelessness. Patrick, a White, male returning volunteer in his early thirties, had spent 4 days and nights volunteering in a Northeastern US homeless shelter as part of his US Catholic schooling. Reflecting on that experience, he said:

> You get there [to the shelter], and you realize it's never as bad as it seems in your head ... [You learn] what a homeless shelter's actually like versus how you read about it in the [newspaper] ... You're like, "People are just people. Like, not everyone in a homeless shelter is there because they're, like, a drug addict." ... You find out some people are there just because of bad circumstances. A lot of them are just – they come over as, like, first-year immigrants, and so there's not an established – you know, they're working below the poverty line anyway; it's why they're at the homeless shelter. They just haven't, like – they don't have any education; they can't get anywhere. (July 11, 2014)

Patrick's comments illustrate the potential role of volunteering in democratic citizenship education. Because volunteering at a local homeless shelter allowed Patrick to make connections with marginalised people, he saw more clearly how structural forces (e.g. access to education, immigration policies) position people differently in matrices of inequality.

In contrast to Patrick's domestic volunteer experience, volunteering at Celebrate Life provided few opportunities for volunteers to learn from or about other people. Because of a language barrier and a lack of structured activities, volunteer tourists rarely interacted with any Celebrate Life residents besides children. Jessica, a White, female, returning volunteer from the US Southeast in her early twenties, explained why this was so. When asked how or whether volunteers were able to develop shared interests with Romanian adults, Jessica said:

> For the adults here, like the caretakers, they don't try to [have intercultural understanding]. They don't talk to us. They get very mad as us. Like, you may have heard already about this, but just a few days ago, we were putting [suntan] lotion on all of the kids because of the sun, and one of the caretakers got very mad and told all the kids to go inside. And she said that all the volunteers load up their kids with lotion and food. We're trying to get them to put sunscreen on so they don't get skin cancer or sunburns! So, I think the adults, like the caretakers, do not try to have any cultural understanding. They just look at us like we're here on vacation and want to take pictures, stuff like that. They don't talk to us, really. But it's also very difficult because there's a language barrier. Most of the adults here do not speak any English. (July 10, 2014)

A significant language barrier did prohibit Romanian adults and volunteers from conversing. Romanian language skills were not a prerequisite for volunteers. Relatedly, many Celebrate Life adult residents had either limited or interrupted formal schooling, which precluded advanced proficiency in English or another world language. That volunteers stayed in a separate house also physically distanced volunteers from adults. There were limited spaces on Celebrate Life's campus in which volunteers and adults could interact or have chance encounters.

Romanian adults at Celebrate Life also admitted to having few interactions with volunteers. Nadia, a thirty-something woman from the local village who had worked numerous years as a caregiver at Celebrate Life, responded as follows when asked about her understanding of volunteer tourists:

Nadia: I don't know too much about what they [volunteer tourists] do. What their purpose for coming is – but they are here to help. With what is needed. I do not know, perhaps they ask the office what – what the needs are – I do not know.
Iordache-Bryant: Have you interacted with any of them?
Nadia: No. (July 10, 2014)

In her study of social justice education in elite US secondary schools, Katy Swalwell (2013) notes that in classrooms where students maintained a sense of *noblesse oblige* toward marginalised groups, '[e]ncouraging students' voice was more prominent than encouraging students to listen ... whether to each other or to other people' (p. 76). Listening across difference is a key component of justice-oriented, democratic citizenship education. However, neither volunteer tourists nor Romanian adults at Celebrate Life seemed to listen to each other. In fact, the few interactions that did occur between them often resulted in tension. Rebecca, a White, female volunteer coordinator from the US Northeast in her mid-twenties, responded to the question, 'How about the community at large? What is their understanding or sense of the volunteers who come and go?' with a story of intercultural exchange gone awry. She said:

It's hard to tell [what locals' perceptions of volunteer tourists are]. Most of them have very little interaction with the volunteers. And the only times that I've really ever seen interaction, it's not in the most positive light. Because usually it happens up at the [local bar], and so then, you know, you're not in [Celebrate Life]; you're right in the village. So it makes it, you know – you're on their turf kind of thing. They've never – I've never seen a villager instigate anything with a volunteer, but if the volunteer instigates something, words can be thrown and all of that. (July 9, 2014)

Instead of allowing disparate groups to come together and explore mutual interests, volunteer tourism at Celebrate Life produced a space of accentuated difference. Global citizenship education, in this context, produced few opportunities for Romanians and volunteers to learn from or about one another.

The volunteer tourists we spoke to, however, seemed untroubled by this lack of connection. They directed their service efforts towards children alone. Gwen, introduced previously, explained:

Honestly, I don't really work with – I don't work with the mothers. And I think simply – that's just simply because ... I have never grew a passion for working with adults. So, I think I've kind of just stayed away from that area. (July 14, 2014)

Volunteer tourism's overall emphasis on children (cf. Sinervo, 2011) may help account for Gwen's stated disinterest. A dominant discourse in volunteer tourism – taken up by our focal NGO on its website – is that children are amenable to outside interventions and should be the focus of volunteer efforts (cf. Butcher & Smith, 2015).

Gwen's lack of interaction with adults, however, had significant perceptual effects. It may account for why she later ascribed the physical violence she observed at Celebrate Life to Romanian culture, not individual adults' troubles. Here, Gwen shares what volunteer tourism has taught her about Romania:

In regards to the culture, I've learned that, you know – how they discipline kids can sometimes, you know, be a [slap sound], you know, a spank or something like that, and when I first saw that, I was like, "Oh my gosh – that's abuse. That is not right, that's not good." But in the Romanian culture I have learned ... that is *in their culture*. It's part of their culture, and it's just how they discipline kids ... This is how it is here. (July 14, 2014, emphasis in original)

Although she rarely interacted with Romanian adults, Gwen asserted that what she witnessed at Celebrate Life was evidence of these adults' culture. In the volunteer tourism gaze – one unmediated by deep understandings of local people and context – material privation, behaviours associated with stress, and signs of social immobility are often explained as cultural diversity (Crossley, 2012). Because the NGO did not formally educate volunteers about the service context, their understandings of it relied upon visible markers of difference, conversations with other volunteers, and personal judgments. This finding coheres with related research, which suggests that global citizenship education 'privileges individual experience and affect but deprioritizes a commitment to knowledge and truth beyond experience' (Kolb cited in Butcher & Smith, 2015, p. 95).

That volunteers conflated what they observed at the intentional community, Celebrate Life, with the country, Romania, also highlights a key aspect of global citizenship education: it relies on a discourse of immersion in other countries that 'perpetuates a view of the world as a mosaic made up of fundamentally different groups that are internally homogeneous' (Doerr, 2012a, p. 16). In both its structural organisation and ideological orientation, Celebrate Life could be considered a subculture in any country. However, many volunteers offered that what they observed there reflected Romania as a whole.

Field notes recorded by the second author capture this synecdoche. In an example of discrepant data, she describes her reassessment of another volunteer (Jessica, introduced previously) when Jessica sought her out to tease apart Celebrate Life, the intentional community, from Romania, the country. She wrote:

> The manner [Jessica] speaks in is very authoritative in any subject – she claims knowledge about Romania, about Celebrate Life, about the U.S. – and she speaks about these topics with a sense of entitlement, of expertise that she may or may not factually have. This manner of hers really put me off her and it is funny how my tonight's change in opinion of her is related to her acknowledgement of the fact that she is not an expert. She started speaking about sexism and she said that she actually wanted to ask me about it – she framed it like this: "because perhaps, it is more of a Celebrate Life thing than a Romanian thing." I appreciated her saying that this place might not reflect Romanian culture and that Romania stands for more than Celebrate Life. I wondered if I had caused this change in attitude – I am very much Romanian, but not the kind of Romanian that lives here [at Celebrate Life]. I stand for maybe something that may be more representative of the population and I may offer insight from a different standpoint on Romanians.

It has been noted that globalist projects rely on a 'direct, mutually dependent co-construction of deterritorialised people (i.e. globalised students) and territorialised people (i.e. parochial hosts), who are placed in a relation of power in a context where the 'global' is valorised in the name of creating global citizens (Doerr, 2012a, p. 4). In other words, globalist projects cast host people as *local* and temporary visitors as *global*. The second author's positionality as a Romanian volunteer being educated in the US, however, destabilised this construction. Jessica used Iordache-Bryant's cosmopolitan presence to broaden her understanding of Romania.

Jessica's reaction to the second author contrasted sharply with those of other volunteers, many of whom seemed to prefer volunteer tourism at Celebrate Life to a more complex experience of Romania and the place of visitors in it. For example, a 16-year old, White, female, Canadian volunteer refused to take advantage of an NGO-organised, weekend afternoon excursion to Braşov, a Transylvanian city known for its lively café culture and baroque architecture. Instead of wandering Braşov with us for the allotted hour and a half of our visit, she stayed in our mode of transport – a van – and looked at pictures on her phone of herself alongside Celebrate Life's resident children.[1] The second author was distressed by this incident, as she wanted volunteers to have a fuller understanding of her country than time at Celebrate Life alone could provide.[2]

Voluntary service as primarily about the self

Findings from this study suggest that participation in volunteer tourism may not provide volunteers with a clear way to be of use to a host community. Nor does volunteer tourism necessarily provide participants with increased knowledge of their host site. Despite these limitations, participants in this study framed volunteer tourism as more valuable than volunteering at home. They cited the benefits of international travel and cultural immersion to make this claim.

Casting volunteer tourism as more exciting than domestic volunteering, Rebecca, introduced previously, stressed its travel component. She said:

> I think it feels cooler to help someone that's outside of your country than it does inside your country because you get to travel.... I volunteered a lot in the U.S. as well, but I still look more fondly on my [previous volunteer] trip to Costa Rica than I do working at the food bank. Because my trip to Costa Rica was exotic – it was intriguing. It was – you know, different. I got to go on a plane; I got to use my passport; I got to buy all these cocoa butter things; I got to eat cacao beans in the middle of the jungle. And so to me, that's way cooler than helping the food bank, even though the food bank is a really honorable thing to do. (July 9, 2014)

Shannon, introduced above, spoke similarly. She explained that participation in volunteer tourism offers a greater challenge than domestic volunteering. In her words:

> The cultural immersion [in volunteer tourism] is real versus doing a soup kitchen every Saturday. And I've done a soup kitchen every Saturday, and it's different, it's great, but you can go home. Like, you just go, but then you're done. So it's different, you know? This is more of a complete immersion. There's so much more to tackle when you're living somewhere and it's 24/7 .... You're never off. [You're] always in the same culture. Even if you're not physically doing something, you're still here living this life, you know? Right? I don't know. It's like you can't – it's not – even if you're in your room, you're still in Romania. (July 9, 2014)

Here, Shannon conflates volunteering with being in Romania. Nowhere else in our data set do we get as a clear a picture of what volunteering has morphed into under the demands of neoliberal governmentality and global citizenship. Primarily a project of shaping the self, volunteering for global citizenship indexes personal resiliency, comfort with difference, and an opportunity to overcome hardship (see Zemach-Bersin, 2009, on discourses of conquest in globalist projects). By comparison, domestic, voluntary civic obligations such as working at a food bank or attending a city council meeting lack adventure and are personally unchallenging.

Some suggest that because volunteer tourism's value is measured in terms of global citizenship, what one actually does on a volunteer tourism project matters little (cf. Vrasti, 2013). We argue, however, that views of volunteering that diminish its purpose and role in domestic settings bear closer scrutiny. Polishing one's church rectory, driving a neighbour to a polling place, or canvassing with a political petition: none of these activities involve complete immersion, except in a cause greater than oneself. To the extent that global citizenship education may hollow out the meaning and significance of *volunteering* – particularly volunteering for democratic citizenship education – more skepticism towards global citizenship's value seems warranted.

Conclusion

We conclude by suggesting that global citizenship education through volunteer tourism has troubling implications for democratic citizenship education. Volunteering as civic engagement at Celebrate Life had severe limitations. It did not meet the community's material needs or teach volunteers about Romania. Nor did it help volunteers understand the intentional community movement – a movement that it is resolutely of-the-moment as more ecovillages, co-housing projects, and continuing care retirement communities arise worldwide (Brown, 2002). Volunteer tourism at Celebrate Life emptied *volunteering* of its social justice potential: to do needed material labour, *gratis*, and simultaneously learn why needs exist. Global citizenship education, in effect, may negatively impact citizenship practices at other scales.

A standard concern in democratic citizenship education is whether volunteering situates privileged people as working *for* rather than *with* the marginalised (cf. Swalwell, 2013). In justice-oriented democratic citizenship education, volunteering can allow people in different social structural locations to come together to address the root causes of problems (Lisman, 1998). In our data set, however, volunteering did not even allow privileged people to work *for* a marginalised community, let alone work *with* them. It also created a false equivalency between qualitatively different kinds of social action: that which is foremost a personal challenge (e.g. living in a different country) and that which primarily benefits others (e.g. feeding hungry neighbours). A comment made by a Romanian Celebrate Life employee to the second author captures this distinction. She said, 'We don't call Romanians "volunteers"', referring to visiting Romanian nationals who donate construction labour or schoolwork help to community members. Among Celebrate Life insiders, the term, *volunteer*, indexed the entrepreneurial, active citizen: one who takes personal risks yet whose service impact is negligible. Global citizenship education, in this context, was closely aligned with neoliberal governmentality (cf. Lyons, Hanley, Wearing, & Neil, 2012).

In the small, Iowa town in which we live(d) and wrote this paper (Iordache-Bryant having graduated), volunteer-based social programmes struggle mightily. The local, no-kill animal shelter relies on a small band of committed volunteers to care for and place homeless dogs and cats. Similarly, the community food bank is stretched to capacity with only a few people collecting and distributing canned goods. Yet, the students who attend the elite liberal arts college situated in this same town are rarely advised to donate their energies to such causes. They are encouraged instead to become entrepreneurial, global citizens by engaging in globalist projects. In fact, the college earns rankings points for students' doing so in numerous metrics of institutional quality. What are the implications of global citizenship, we ask, when domestic service programmes' volunteer needs go unmet? How is the concept of *volunteering* being altered because of its association with tourism and global citizenship?

Notes

1. Although on-site NGO staff members specifically asked volunteers not to take pictures of Celebrate Life's residents (in part because many of them were in hiding from domestic violence), many did so anyway. We also saw evidence that volunteers were posting these pictures to online social media sites like Facebook.
2. To this end, the second author made sure that the first author visited Bucharest, the famed Transfăgărășan highway, and a well-known Soviet-era salt mine prior to her return to the US.

Disclosure statement

No potential conflict of interest was reported by the authors.

References

Binkley, S. (2007). Governmentality and lifestyle studies. *Sociology Compass, 1*(1), 111–126.

Brown, S. L. (Ed.). (2002). *Intentional community: An anthropological perspective*. Albany, NY: State University of New York Press.

Burchell, G., Gordon, C., & Miller, P. (Eds.). (1991). *The Foucault effect: Studies in governmentality*. Chicago, IL: University of Chicago Press.

Bush, G. H. W. (1988, August 18). Address accepting the presidential nomination at the Republican National Convention in New Orleans. The American Presidency Project. G. Peters and J.T. Woolley. Retrieved from http://www.presidency.ucsb.edu/ws/?pid=25955

Butcher, J., & Smith, P. (2010). "Making a difference": Volunteer tourism and development. *Tourism Recreation Research, 35*(1), 27–36.

Butcher, J., & Smith, P. (2015). *Volunteer tourism: The lifestyle politics of international development*. New York, NY: Routledge.

Charmaz, K. (2006). *Constructing grounded theory: A practical guide through qualitative analysis*. Los Angeles, CA: Sage.

Conran, M. (2011). They really love me! intimacy in the volunteer tourism encounter. *Annals of Tourism Research, 38*(4), 1454–1473.

Cross-Cultural Solutions. (2014). Our Philosophy. Retrieved from https://www.crossculturalsolutions.org/our-philosophy

Crossley, É. (2012). Poor but happy: Volunteer tourists' encounters with poverty. *Tourism Geographies: An International Journal of Space, Place and Environment, 14*(2), 235–253.

Dean, M. (1999/2010). *Governmentality: Power and rule in modern society* (2nd ed.). Los Angeles, CA: Sage.

Doerr, N. M. (2012a). Do 'global citizens' need the parochial cultural other? Discourse of immersion in study abroad and learning-by-doing. *Compare, 43*(2), 1–20.

Doerr, N. M. (2012b). Study abroad as 'adventure': Globalist construction of host-home hierarchy and governed adventurer subjects. *Critical Discourse Studies, 9*(3), 1–12.

Emerson, R. M., Fretz, R. I., & Shaw, L. L. (1995). *Writing ethnographic fieldnotes*. Chicago, IL: The University of Chicago Press.

Global Service Corps. (2008). *Global Service Corps*. Retrieved from http://www.globalservicecorps.org

Gray, N., & Campbell, L. M. (2007). A decommodified experience? Exploring aesthetic, economic, and ethical values for volunteer ecotourism in Costa Rica. *Journal of Sustainable Tourism, 15*, 463–482.

Guiney, T. (2017). Orphanage tourism and development in Cambodia: A mobilities approach. In J. Rickly, K. Hannam, & M. Mostafanezhad (Eds.), *Tourism and leisure mobilities: Politics, work, and play* (pp. 178–192). New York, NY: Routledge.

Heath, S. (2007). Widening the gap: Pre-university gap years and the economy of experience. *British Journal of Sociology of Education, 28*(1), 89–103.

H.R.1388 - Serve America Act. (2009). 111th Congress (2009-2010). Retrieved from https://www.congress.gov/bill/111th-congress/house-bill/1388

Inda, J. X. (Ed.). (2005). *Anthropologies of modernity: Foucault, governmentality, and life politics*. Malden, MA: Blackwell Publishing.

Jakubiak, C. (2015). A pedagogy of enthusiasm: A critical view of English-language voluntourism. In J. A. Álvarez Valencia, C. Amanti, S. Keyl, & E. Mackinney (Eds.), *Critical views on teaching and learning English around the globe: Qualitative research approaches* (pp. 193–209). Charlotte, NC: Information Age Publishing, Inc.

Jakubiak, C. (2017). Mobility for all through English-language voluntourism. In J. Rickly, K. Hannam, & M. Mostafanezhad (Eds.), *Tourism and leisure mobilities: Politics, work, and play* (pp. 193–207). New York, NY: Routledge.

Kamau, L. J. (2002). Liminality, communitas, charisma, and community. In S.L. Brown (Ed.), *Intentional community: An anthropological perspective* (pp. 17–40). Albany, NY: State University of New York Press.

Keese, J. R. (2011). The geography of volunteer tourism: Place matters. *Tourism Geographies, 2*, 257–279.

King, S. (2006). *Pink ribbons, inc.: Breast cancer and the politics of philanthropy*. Minneapolis, MN: University of Minnesota Press.

Lisman, C. D. (1998). *Toward a civil society: Civic literacy and service learning*. Westport, CT: Bergin & Garvey.

Lyons, K., Hanley, J., Wearing, S., & Neil, J. (2012). Gap year volunteer tourism: Myths of global citizenship? *Annals of Tourism Research, 39*(1), 361–378.

McBride, A., & Sherraden, M. (Eds.). (2007). *Civic service worldwide: Impacts and inquiry*. Armonk, NY: M.E. Sharpe.

McGehee, N. G., & Andereck, K. (2009). Volunteer tourism and the "voluntoured": The case of Tijuana, Mexico. *Journal of Sustainable Tourism, 17*(1), 39–51.

Obama, B. (2009). United we serve: Remarks of President Obama. Retrieved from http://www.serve.gov/?q=site-page/remarks

Perlstein, D. (2008). Freedom, liberation, accommodation: Politics and pedagogy in SNCC and the black panther party. In C. M. Payne & C. S. Strickland (Eds.), *Teach freedom: Education for liberation in the African-American tradition* (pp. 75–94). New York, NY: Teachers College Press.

Rose, N. (1999). *Powers of freedom: Reframing political thought*. Cambridge: Cambridge University Press.

Schwartz, K. C. H. (2010). Unequal opportunities for citizenship learning? Diverse student experiences completing Ontario's community involvement requirement. (*Unpublished master's thesis*). The University of Toronto, Toronto, Ontario, Canada.

Simpson, K. (2004). 'Doing development': The gap year, volunteer tourists and a popular practice of development. *Journal of International Development, 16*, 681–692.

Sinervo, A. (2011). Connection and disillusion: The moral economy of volunteer tourism in Cusco, Peru. *Childhoods Today, 5*(2), 1–23.

Swalwell, K. (2013). *Educating activist allies: Social justice pedagogy with the suburban and urban elite*. New York, NY: Routledge.

The Foundation for a Better Life. (2014). Pass it on: Billboards. Retrieved from http://www.values.com/inspirational-saying-billboards

Tsing, A. L. (2000). The global situation. *Cultural Anthropology, 15*(3), 327–360.

United Planet. (2016). 15 years of building a community beyond borders. Retrieved from http://www.unitedplanet.org

Vrasti, W. (2013). *Volunteer tourism: Giving back in neoliberal times*. New York, NY: Routledge.

Wearing, S. (2001). *Volunteer tourism: Experiences that make a difference*. Oxon: CABI Publishing.

Westheimer, J., & Kahne, J. (2004). What kind of citizen? The politics of educating for democracy. *American Educational Research Journal, 41*(2), 237–269.

WorldTeach. (n.d.). WorldTeach. Retrieved from www.worldteach.org

Zemach-Bersin, T. (2009). Selling the world: Study abroad marketing and the privatization of global citizenship. In R. Lewin (Ed.), *The handbook of practice and research in study abroad: Higher education and the quest for global citizenship* (pp. 303–320). New York, NY: Routledge.

Mediating global citizenry: a study of facilitator-educators at an Australian university

Tamara Young, Joanne Hanley and Kevin Daniel Lyons

ABSTRACT
Research concerned with global citizenry in tourism is often focused on the potential outcomes of educational travel on the global competencies of student travellers. The experiential learning of educational travel – increasingly represented by short-term, faculty-led programmes – can be transformative for students. It is argued that educational travel shifts the focus of young people from a narrow or self-oriented position, to a broader and more encompassing global perspective to become socially aware and responsible global citizens. Whilst such outcomes of educational travel are evidenced in academic literature, an aspect of educational travel that has been overlooked concerns the mediation of global citizenship. This qualitative study investigates the role played by university staff who facilitate short-term educational travel, finding that, in addition to playing a key role in curriculum delivery and programme logistics, university staff are active in global citizenship education through the mediation of student experiences in unfamiliar cultural contexts. The role of academic staff in educational travel can be compared, theoretically, to that of a tour guide. Educator-facilitators are central to developing the global competencies of students and nurturing global citizenship.

Introduction

A primary mechanism for implementation of policy on global education has been investment in outbound mobility programmes that enable university students to engage in international educational travel. The rationale for government and university-sponsored educational travel, which exists in various forms, is that educational travel can yield substantial improvements in the global competencies of students in a relatively short period of time. Evidence suggests that the outcomes of international mobility for students include: personal development (Harrison, 2006), professional development (Crossman & Clarke, 2010; Lyons, Hanley, Wearing, & Neil, 2012), acquiring functional knowledge (McKeown, 2009; Pitman, Broomhall, McEwan, & Majocha, 2010), developing intercultural competencies (Hovland, 2009; Williams, 2005), bridging cultural distance (van 't Klooster, van Wijk, Go, & van Rekom, 2008), promoting awareness of global issues (Chieffo & Griffiths, 2004; Dolby, 2007; Hendershot & Sparandio, 2009), and nurturing global citizenship (Stoner et al., 2014; Tarrant et al., 2011).

For university students engaging in structured and purposeful educational travel, the transformative impacts of their international experiences have been linked to fateful (Giddens, 1991) and critical moments (Holland & Thomson, 2009; Thomson, Bell, Holland, Henderson, McGrellis, & Sharpe, 2002), whereby experiential learning whilst abroad can be identity forming and/or identity changing (Desforges, 2000; Noy, 2004; Wearing, Stevenson, & Young, 2010). The effect of such experiences has been found to shift the focus of young people from a narrow or self-oriented position, to a broader and more encompassing global perspective (Stoner et al., 2014). This shift, it can be argued, is an essential step for young people to become socially responsible and culturally aware global citizens.

According to Stoner et al. (2014, p. 151), educational travel programmes 'can serve to create a transformative educative experience where students reconsider and reshape fundamental issues from a global perspective'. Due to the assumed efficiencies and practicalities of short-term programmes and their impact on young people, there has been a shift away from traditional semester exchange study abroad programmes to short-term outbound mobility programmes. Short-term mobility programmes (STMPs) vary in length from a week to several weeks, and include an array of faculty-led learning activities, including, study and cultural immersion tours, international work integrated learning experiences, volunteering and service learning. Advocates of educational travel argue that, regardless of the length of the mobility programme, a well thought out and professionally led programme that is coupled with a sound pedagogical framework will promote learning outcomes that go beyond the impacts of traditional campus-based instruction (Chieffo & Griffiths, 2004; McKeown, 2009; McLaughlin & Johnson, 2006; Stone & Petrick, 2013; Tarrant et al., 2011).

With this increase in faculty-led programmes aimed at creating global citizens, the role that university staff play in facilitating and mediating student experiences abroad has not been researched. In addition to curriculum development and delivery and the management of programme logistics, individual university staff are travelling overseas with students to guide learning, negotiate experiences, and perform pastoral care responsibilities (Young & Lyons, 2016). Whilst studies of educational tourism are increasingly commonplace, they are most often outcome-oriented with research directed to the benefits of short-term programmes on student participants.

An area of educational travel that has received little scholarly attention is that educational travel, as with all tourism experiences, is a mediated activity (Wearing et al., 2010). The student involved in educational travel is essentially a tourist in a structured learning programme, and the STMP facilitator-educator is, in essence, a tour guide taking on the responsibilities of 'leader' and 'mediator' (Cohen, 1985, 2004) of the tourist experience. Tour guides are ever present in tourism interfacing between traveller and travelled cultures (Jennings & Weiler, 2006; Young, 2009; Young & Lyons, 2011) and this study examines the role of facilitator-educators of STMPs in nurturing global citizenship as an outcome of educational travel. Thus, our unique contribution to this special issue on tourism, cosmopolitanism, and global citizenship, is our focus on the important and often taken for granted role that university staff play in imparting knowledge for cultural learning and in the mediation of global citizenry.

Educational travel, cultural competency, and global citizenship

The view that global travel is a conduit for fostering globally competent young people is not a new idea, and one that existed before the advent of government and university-sponsored educational travel programmes. Travel has long-been recognised as a 'rite of passage' for young people (Wearing et al., 2010), and is often endorsed as a transformative experience in building cross-cultural understanding, an appreciation of diversity, and promoting global awareness (Lyons et al., 2012; Wearing et al., 2010). However, there has been debate in tourism research about whether such outcomes are an innate product of the act of travelling overseas, or whether they result from carefully honed and developed products and services offered by the tourism industry (e.g. organised tours and volunteer tourism organisations) and more recently universities (e.g. Study Abroad and STMPs).

It is commonplace in the tourism literature to suggest that overseas travel provides an important context for informal learning opportunities, as well as for international and cross-cultural understanding (Karshenas, 1996; Ketabi, 1996; O'Reilly, 2006). Travel provides a means for the accumulation of experiential knowledge (Desforges, 2000; Wearing et al., 2010), a knowledge that contrasts with that of formal education. Some forms of tourism (particularly, educational travel) are recognised as very important informal contexts with the potential for transformative lifelong learning (Pitman et al., 2010). Supporters of this perspective argue that tourism can function as an important contributor to the development of global citizenship, through international and cross-cultural understanding, the sharing of values, mutual support, disabusing of stereotypes, and exchange of value (Ketabi, 1996; Lyons et al., 2012).

At an individual level, global citizenship has been defined as a 'meritorious viewpoint that suggests that global forms of belonging, responsibility, and political action counter the intolerance and ignorance that more provincial and parochial forms of citizenship encourage' (Lyons et al., 2012, p. 361). From a theoretical perspective, global competency has been operationalised in a variety of ways. For example, global competencies are important, especially in relation to an individual's capacity to adapt to global changes that affect career trajectories. Lyons (2010) notes that experiential learning, often associated with global educational travel, is a means by which individuals can develop competencies to strategically craft adaptable careers in relation to opportunities that the changing global context offers.

Not all outcomes of global educational travel, however, are explicitly related to theoretical constructs like global citizenship or career adaptability. For instance, an increase in academic achievement has been tied to participation in various forms of global educational travel (Sutton & Rubin, 2004). A social identity-based theory of global citizenship (Hogg, 2006) is another way of understanding the potential learning outcomes of global educational travel. Reysen and Katzarska-Miller (2013) argue that global educational travel leads to changes in constructions of the self through the internalisation of a global identity, and the subsequent development of pro-social values, attitudes, and behaviours.

In practice, the experiential learning that takes place through educational travel is likely to be carefully managed by the university staff running the learning programme. Student involvement and immersion in the host country, therefore, is undeniably mediated. Indeed, direct experience with places, people, and culture made possible through international travel are always mediated through various 'experiential sources' of information (Dann, 1996), including verbal communication at destinations between 'guests' and 'hosts', between tourists themselves, and, for the purpose of this paper, between university staff and students. Thus, the experiences of educational tourists are mediated through various 'gatekeepers' (du Cros & McKercher, 2015) of information-providing intermediaries situated between the traveller and their travelled destination.

As argued above, the roles performed by university staff can be compared to that of tour guides, given their responsibility for leading and mediating the student experience. Cohen (2004) argues that the role of the modern tour guide has shifted away from being primarily logistical to a focus on the interactional and communicative mediation of the tourist experience. Tour guides as mediators, whilst still involved in the instrumental and social components of leadership, ultimately perform a communicative role that incorporates the selection, provision, interpretation, and fabrication of information (Cohen, 2004).

Tour guide as mediator

Using Cohen's (1985, 2004) seminal framework on tour guiding, we conceptualize facilitator-educators as tour guides, whose meditation role involves a range of communicative responsibilities incorporating selection, information, and interpretation. Selection refers to the role of the guide in pointing out 'objects of interest' that are selected and deemed worthy of the attention of tourists. In this sense the tourist experience is manipulated, in that objects of interest are selected and tourists are shown what the guide wants them to see and learn. A tour guide is, therefore, the gatekeeper of information and disseminator of knowledge. However, that the information imparted is rarely neutral and the information provided may engender wider social, cultural, or political implications that may influence tourists' impressions and attitudes. The process of interpretation is one of the key areas whereby guides mediate encounters between cultures – that is, between the travelling culture and the travelled culture – and, in this sense, the guide acts as 'culture-broker' (Cohen, 2004). In addition to performing these mediation responsibilities, the educator-facilitator's role is to maximise the learning experience in professional contexts of their discipline at the same time as ensuring global citizenry as a learning outcome of educational travel.

Educational travel: the Australian context

The short-term mobility of Australian university students has proliferated over the past 10 years and is continuing to rapidly grow (Harrison & Potts, 2016). Harrison and Potts (2016, p. 4) attribute this increased student engagement in learning abroad to 'a major cultural change' including increased awareness to the benefits of learning abroad, the increasingly competitive global employment market, and systemic change to the policies and practices supporting student participation in outbound mobility programmes. Increased student engagement in learning abroad programmes, such as STMPs, is made possible by federal government funding initiatives aimed at increasing the international experiences of students across all fields of tertiary study. The Australian Government's National Strategy for International Education (2016, p. 4) recognises that 'Australian international education is a core element of Australia's economic prosperity, social advancement and international standing'.

Federally funded initiatives include: *The New Colombo Plan* which focuses on undergraduates studying and undertaking internships in the Indo-Pacific region, and *Endeavour Scholarships and Fellowships* which supports Australians to partake in study, research, or professional development across the globe. *New Colombo*, which uses the slogan 'Connect to Australia's future – study in the region' (Department of Foreign Affairs and Trade, 2016), is expanding from a pilot phase to full implementation over the next few years, with A$160 million committed to enable more than 10,000 Australian students to participate in educational travel by 2019. Australian initiatives echo a similar policy approach in the United States, where the bi-partisan Lincoln Commission set a target of one million students undertaking global educational travel by 2016. The US initiative recognises that 'what nations don't know can hurt them … the stakes involved in study abroad are that simple. For their own future and that of the nation, it is essential that college graduates today become globally competent' (The Lincoln Commission, 2005, p. 3).

In response to government policy on expanding global education through outbound travel, universities throughout Australia are investing in a range of outbound mobility programmes. The overarching aim of global education opportunities available to Australian university students (including semester exchange, intensive short courses, international work placements, and cultural immersion tours) is, expressly, to create global citizens who are socially responsible, globally aware, and committed to civic engagement (Department of Education and Training, 2016; Department of Foreign Affairs, 2016).

Since the first major data collection on Australian student outbound mobility programme participation rates was undertaken in 2005, the percentage of undergraduate students participating in these programmes had more than doubled by 2012 (Olsen, 2013). The most recent published data on Australian university student participation in learning abroad experiences reports that approximately 32,000 students participated in an outbound mobility programme in 2014 (Harrison & Potts, 2016). Short-term programmes (i.e. experiences that are two weeks to under a semester in length) represented 56% of all outbound global mobility programmes in 2014 (Harrison & Potts, 2016). The majority of Australian universities have strategic plans identifying outbound mobility as central to realising the goal of producing globally competent graduates, with the meeting of participation targets generally cited as the key indicator that such goals are being achieved.

For example, at the University of Newcastle (UON), the regional Australian university where this case study for this research took place, a key focus of the current strategic plan is to position itself as an international leader in outbound student mobility (University of Newcastle, 2015). The strategy entitled *Global Mobility and Employability* states:

> We will increase outbound student mobility across all levels, on semester exchange, WIL, internship, research experiences, short courses and study tours, with streamlined, structured and user friendly information. We will develop productive networks with international Alumni, Austrade, Governments, research institutes, businesses, and professional groups and universities to support a high quality student experience in country. (University of Newcastle, 2015, p. 17)

Given that this is a lead strategy for UON, this university provides a rich setting for investigating the mechanism of mediating global citizenry through outbound student mobility, including STMPs. Some undergraduate and postgraduate programmes at UON now offer elective courses that incorporate an international study experience, and each one of its faculties has been successful in gaining government funding through initiatives including *New Colombo* and *Endeavour mobility* grants.

Methodology

As highlighted in the literature review, extant research is typically focused on the learning and developmental outcomes of educational travel for students, which includes the development of global competencies and the nurturing of global citizenship. Our study addresses an alternative viewpoint in that it focuses on the role of university staff in the mediation of these outcomes. In a recent study on global education programmes, Wearing, Tarrant, Schweinsberg, Lyons, and Stoner (2015) argue that a structured learning context is necessary for travellers to experience, grapple with, reframe, and reflect on issues global in nature. Clearly, being a global citizen has become the overarching goal and outcome of educational travel, which leads to the research questions we address in this paper. These included the following: Why do academic staff become involved in the facilitation of STMPs? What does global citizenry mean to those academic staff who facilitate STMPs? Are those staff committed to the goal of global citizenship as an outcome of their programmes? How do staff develop and nurture global citizenship of students involved in their programmes? These questions together inform an overarching question concerned with the ever presence of mediation in tourism and the role of facilitator-educators as an interface between students and their experience of other cultures and societies: How is global citizenship mediated by university staff to create cosmopolitan and culturally competent students?

This is an exploratory study, its purpose being to address an existing topic from a new perspective (Mason, Augustyn, & Seakhoa-King, 2010), as detailed above. The qualitative study presented in this paper focuses on academic and professional staff at UON who were involved in the facilitation of government-funded outbound student mobility programmes. Participants were recruited through UON Global, the international relations office responsible for the administration and coordination of all university study abroad and exchange programmes. This resulted in a sample of 12 participants who had facilitated a variety of discipline-specific mobility programmes.

Ten participants were academic staff representing seven disciplines across two faculties: The Faculty of Business and Law and The Faculty of Education and Arts. The disciplines included: law and legal studies (Newcastle Law School); leisure and tourism studies, accounting and finance, international business and management (Newcastle Business School); speech pathology (School of Humanities and Social Science); education (School of Education); and, fine art (School of Creative Arts). Two participants were professional staff directly involved in managing outbound student mobility, one at a university-wide strategic level and the other faculty-based. The types of STMPs facilitated by the staff included cultural immersion tours and international work placements. Some participants had facilitated multiple trips over a number of years. Countries visited included a broad range of destinations across the globe.

We conducted semi-structured interviews with all participants. Two staff were interviewed more than once, with the aim of capturing their pre-departure and post-trip understanding of their roles as STMP facilitator. Each interview lasted between 45 and 90 minutes and was recorded and later transcribed. Participants in the study were assigned pseudonyms at the time of transcription to protect their anonymity in the reporting of the research findings. As this study involves new theoretical ground, an inductive research approach framed the data collection to generate key themes and concepts. Given the exploratory nature of the study, the findings discussed below are the initial findings of a broader study concerned with the roles of university staff in STMPs. A limitation of this study is the one-dimensional and single institution case study. However, this study provides important initial findings that lay a rigorous empirical base for future mixed method studies on the role of facilitator-educators in creating global citizens.

Findings

In describing their role as STMP facilitator-educators, the study participants demonstrated that there are multiple ways of describing global citizenship and multiples ways of viewing their role in creating global citizens. Our analysis of the data provided three overarching themes that frame the presentation of our findings, including meanings of global citizenship, nurturing global citizens, and mediating global citizenry.

Meanings of global citizenship

It is clear that study participants recognised their role in nurturing the global citizenship of students. However, their understanding of what a global citizen means was rarely explicit. Despite probing during the interviews to elicit definitions from participants, global citizenship was more often explained as experiential. Some explanations, as outlined below, are helpful in coming to terms with how our participants understood this complex term. The following phrases provide a snapshot of commonly expressed understandings of the meaning of the term: 'thinking global', 'a globalised perspective', 'globalised learning', 'working in a globalised world', and 'creating connected individuals in a global sense'.

One facilitator-educator regarded global citizenry as the opposite of ethnocentrism, with the goal being his students 'came out of their Australia-centric, US-centric, social media and just realise that there is an entirely different world out there in [which] people live entirely different lifestyles compared to the advantageous lifestyle that we have in Australia'. He went on to describe his cultural immersion tour with students as more about recognising their privilege, made possible by students attending classes at an African university where students were:

> Going to classes in universities where there were large numbers of students. The only thing in these lecture halls – which themselves were very, very plain and often, you know, dirty – were just concrete floor, concrete walls covered with mud, metal seats, teacher standing up the front, no computer, no PowerPoints, no nothing. (Daniel)

In this instance, it was the student's capacity to compare the university experience – to 'benchmark' it against their own experiences at an Australian university – that developed their ability to become global citizens. He also suggested that funding Australian students to travel on STMPs is 'global citizenry in action' for the following reasons:

> One, showing the other country that we're interested in relationships and we care about them. And the second thing is, investing in those students because none of those students on that trip will ever forget Africa, east Africa, Kenya and the Kenyan people. Now, if they get to a position in large organisations where they've got opportunities to expand resources into those sorts of areas. (Daniel)

When explaining some of the fears held by students on a South Pacific island work placement visit, another participant described what is best framed as an enlightened version of global citizenry, one that involves being both respectful to the host peoples and conscious of the need for reciprocity:

> [I]t was an amazing, kind of, learning opportunity. [I was] there to be supportive and to explain to them, that this is not Australia and this is a different place where [the people] are very sophisticated and savvy, but they are very short on resources. And they need help. They think that you're going to have help to give. They're not idiots, they're not just going to blindly take whatever help that you offer and just say, 'Oh, you're a smart Australian, we're just going to do that.' But at the same time, they welcome some outside perspective and just additional sets of eyes and to look at things and talk about things. So, believe it or not, this may be shocking, but you can make a contribution. They all said, 'Okay, we'll try, we'll try.' (Stephen)

Thus, an important role of the facilitator-educator as expressed here is to point out to students' respect and reciprocity. That is, that genuine global citizenry is about being on equal footing and not just people from the developed world arriving in the developing world to 'help' or research or exploit.

When asked to explain the meaning of global citizenry and how it related to students, one STMP facilitator-educator saw it very much in relation to the profession:

> In the context of speech pathology ... students [become] aware of how speech pathology as a profession operates internationally. So they would be very aware that in some countries speech pathology is a very established role, and certainly in other countries, there are no speech pathologists, and there's certainly no training programmes ... [They become] aware of the need for speech pathology services internationally and ... then that they could actually then work globally ... So they're aware obviously about ... issues around working with children and adults with cultural awareness diversity ... and would have that respect. (Elizabeth)

Another facilitator-educator, who led a cultural immersion tour in Africa, explicitly discussed the theories she engaged to educate the students about their cosmopolitan identity and how being mindful of their privileged position and 'how they actually fit into this whole system ... [is] where I think they learnt more about themselves'. To educate the students for this deep understanding, in pre-departure lectures, the facilitator-educator:

> talked about them being agents of soft power and why this would be so and that when they go over there, to reflect on what kind of role they were playing as agents of soft power. So they were already mindful of these things and they did embrace them and so, when they did travel to [Africa], I think they were very cognisant of all these things. They were mindful of their position. They were mindful of the relationships of power. (Penelope)

This facilitator-educator went on to explain how being made mindful of soft power helped the students to gain from the STMP experience, and this was evident in an assessment task when 'a few of them in their presentations expressly talked about global citizenry and how they felt themselves as having grown'.

What this suggests is that global citizenry does not just happen, it requires a recognition of one's status as a global citizen – a status based on actual culturally engaged experiences outside of one's home country – something that requires a reflexiveness and reflectiveness on how one might have changed as a result. Which leads to the question of how can we determine whether the objective of creating global citizens takes place on STMPs.

Nurturing global citizens

The study participants recognised that a key purpose of STMPs is the nurturing of global citizens through international travel. At UON, the process for applying for international mobility grants – including choice of destination and content of programme – is by individual initiative guided only by the objectives of the specific grant. For instance, a key goal of the government's funding for student mobility schemes is building global awareness and cultural competencies, with internships through the New Colombo Plan intended to 'provide students with opportunities to test their skills in real life situations, build cross-cultural competencies and develop professional networks that can last a lifetime' (Department of Foreign Affairs, 2016). Whilst individual applications were not analysed for the study, participants were asked to explain how they understand their programme within such broader policy goals.

A number of reasons were provided for staff involvement in STMPs. For the academics, perhaps not surprisingly, the overarching goal of any STMP was a learning activity that enables students to experience other cultures 'first hand' and, as such, can lead to personal enlightenment. As one facilitator-educator notes, educating students through cultural immersions 'opens their eyes to international diplomacy and international global networks and issues'. Indeed, the global citizenship value of STMPs was raised by most participants when they discussed their reasons for being involved. Developing a cultural and worldly identity was oft-cited rationale:

> My reason for applying for these ... is my passion for travel in young people's lives and the significant role that travel does play in identity formation and just becoming ... more worldly and understanding other cultures and other places. (Amanda)

As was developing cross-cultural competencies:

> We want them to be better able to operate cross-culturally to everything from just very simple things about body language and stuff through to different ways of thinking, understanding different political systems and different legal systems and how they operate and different histories and how they shape different countries and cultures. (Nathan)

Recognition that Australian students, particularly those based at a regional university, may lack global awareness – that students may be 'a little bit naïve and inward looking' (Susie) – was also highlighted:

> I just see it as a fantastic opportunity for students to be able to get out of Newcastle and to get out of Australia and to become, if possible, more globally aware citizens and socially responsible citizens. (Amanda)

Some participants discussed the specific demographic make-up of the university whereby many students are 'first-in-family' to attend university but also in terms of their overseas travel experiences:

> From an educational perspective I think it's really important ... Last year and this year I had students who had never travelled internationally before and they had to get passports. (Anna)

For one of the facilitator-educators we spoke to, it is precisely because of the limited international travel previously undertaken by many of the students involved in STMPs that means that global citizenship is being nurtured:

> We had not realised the number of first in family students ... These are kids that have never travelled in their own family's lives, their families have never travelled, they've never had a passport. So that is extraordinary, and the life experience and the change in them they've expressed most amazingly. They just never knew their own capabilities ... To just find their way to an airport, to get on a plane, and to go somewhere particular into countries where English is a second language. So there is their own sense of their accomplishment, but [also] their own sense of what they're capable of learning and doing on their own. (Melanie)

Subsequently, some participants saw opportunities for the enhancement of employability because of increased cultural knowledge and improved international networks:

> [STMP initiatives] can ensure that our students have an understanding that they're working within a globalised world so that's important, that breaks down barriers. (Susie)

> [Students in Africa] were talking to people that worked at the UN, talking to people that worked in different places. They found out the importance of networking to be able to work in one of these global international institutions ... You know, and to look at employment on those sorts of level. (Penelope)

For some study participants, the outcomes of student travel programmes had a more intrinsic and personal dimension meaning that their desire to apply for funding and to facilitate programmes could be used for their own purposes, such as, for improving their teaching and learning activities. For example, one interviewee who has facilitated multiple STMPs described how her involvement in these trips informs her teaching:

> [STMPs] changed the way I deliver my classes I think because I'm wiser and more experienced ... I can use examples and the best thing actually, the most amazing thing, is the compare and contrast ... The fact that you can look back at your own legal system and see its faults because it's certainly not perfect and then you can – and work within another legal system and see there's opportunities for incredible change there. But there's also things we can learn from them as well which is great. I think that's the most important thing actually. (Susie)

Whereas, a clinical educator who has travelled with students to a number of South East Asian countries stated:

> I mean I think initially the highlight is obviously visiting a country that I hadn't visited before, and then obviously getting to know that country well in terms of our profession, speech pathology. I'm looking at what resources they've got. I think just the satisfaction that you're actually providing communication to children that previously may not have had access to communication. I think working with some of the stakeholders, so, certainly in the schools, working with the teachers and the other staff, and you know, exploring with them, what they know and what they don't know. So some training we would certainly do. (Elizabeth)

Other participants described how their involvement in STMPs improved their knowledge and skills as academics, including informing their research interests:

> For me ... involvement in STMP facilitation fits in with my broader research interests and research passions about youth travel. (Amanda)

> I'm a cultural researcher ... and China is one of the main areas that I've been exploring ... So actually running these [STMPs] was ... a win/win, ... Made more connections, able to do more research. (Trevor)

Therefore, STMPs can be regarded as helping academics to build and maintain connections in the other countries. Anna stated about her trips to Zimbabwe, 'the personal rationale is that I had connections there. I know the area and I know that there's work that we can contribute'. In some instances academic staff had a deep knowledge of, and professional connections to, the destination.

The implication here is that well-travelled facilitator-educators with good global connections are well positioned as mediators of global citizenry. However, this is not always the case where staff are themselves new to a foreign cultural setting. In comparison to the cognisance of some participants with the destination countries, other well-travelled interviewees were visiting countries new to them and they expressed how the trips played a role in developing their own global citizenry:

> I've been to most south-eastern Asian countries but [Africa] was still a life changing trip to me. Something that I won't forget. Particularly because we were off the tourist trip where we're in with the people's daily lives and the students and going to [an African] university. (Daniel)

> Because I hadn't been there before, I guess I had my own set of, I don't know, expectations and not knowingness about what the experience was going to be like ... [I'd] done a bit of research and I know my stuff in terms of – I don't know actually, I don't know if I did know my stuff. I did not have any idea what it was going to be like in China. (Amanda)

The above quotes suggest that despite being well travelled, these educator-facilitators had limited cultural knowledge of the destination. Therefore, there are questions around the cultural competence of the facilitators if the purpose of staff travelling with students is to educate them about a particular culture and the implications for their global citizenry. Indeed, such a gap in the cultural competence of some Australian university staff was described as the key motivator for one interviewee's involvement in STMPs. Although this educator has gained substantial funding and organised numerous programmes, she has never personally travelled with students to a destination country. Her main objective is to raise the tolerance and cultural competency of fellow staff:

> I'm very keen to actually build competencies and capacities in other staff. Because one of the biggest problems I think for how international students are treated when they come here is that you have staff that are very intolerant and don't understand what it is to be a student in a new culture. And so I've spent years and years with teachers, with staff in at my office saying, 'I can't teach these students, they don't have good enough English' or 'I can't read their essays', or just being disruptive in that process of understanding. So I get the money, I do the lead work, I justify the whole thing, and then I say to staff 'You're going to take the trip'. (Melanie)

The facilitator-educators we spoke to can be considered trailblazers in the STMP space at UON, as 'early adopters' (Linda) of the government initiatives at this institution. Despite the university being one of many Australian institutions in which student mobility is promoted, it is not common for academic staff to commit time to developing and running programmes that enhance learning outcomes related to creating culturally aware global citizens. Indeed, Linda noted that it has been very difficult to encourage staff to facilitate STMPs, and the feedback she has received from academic staff is that they are not willing to be involved in teaching activities that are time and resource intensive creating an inherent tension with the more pressing need to focus on activities related to research output. But for those staff who are willing to commit to student mobility, there were various ways by which they endeavoured to encourage global citizenship as an outcome of their programmes.

Mediating global citizenry

Beyond understanding that their role is to nurture global citizenship, there were examples expressed by study participants of the ways by which they mediate global citizenry. Participants explained how their programmes were structured and theoretically informed, and how experiences on the ground were managed through such techniques as daily briefings, structured lectures, immersive activities and, on their return, assessable reflective journals and presentations. Secondary to these were observations by facilitator-educators on how students were managing their experiences.

One participant described global citizenship as central to the content of her mobility programme which focused on resources management:

> I was very careful to include into the curriculum everything that I thought that could be achieved in advancing global citizenry to the students as an outcome and in doing that, they had to read all the media reports from release from the Australian High Commission for the previous three years ... [this along with the immersion led to the students having] ... a deeper understanding of the other role diplomacy holds with the relationship between Australia and Kenya. (Penelope)

Another facilitator-educator explained how he helped to shape their experience:

> [A] lot of the debriefing and the making sense of what they were going through, was hugely important. That's certainly one of the things I learned. You can't just assume that because they're there, that they're going to learn all the lessons that we hope they're going to learn. It requires a lot of helping them to see what they're seeing, or to makes sense of what they're seeing. And so, being together more, made that possible. I insisted, essentially every evening, we met for a few minutes. Often for an hour or two. (Stephen)

He explained his method further:

> We met every evening ... 'Okay, what issues are we having? What's going on?' Sometimes they would say them. Sometimes I would ask other questions and I could derive from what they were saying. 'What about this? Is this a problem? Is it, you're in this different culture and different environment?' (Stephen)

This approach was echoed by an academic in the school of education who has taken students to visit schools in southern Africa on two occasions. She talks about 'the hard work around dealing with white guilt'. Particularly for Australian students with no experience of developing nations:

> I explain what's going to happen. I explain the tour and what they're going to confront and how hard it might for them on some days and the amount of work they have to do. We talk about volunteerism and the dangers of thinking you're going to save the world and try and keep their expectations in a box. Then I mean from one of the schools we go to the children don't have a feeding program beyond a bit of porridge and they don't have a deworming program which means that there's a lot of malnutrition and it's really quite obvious ... A lot of the kids will be orphans and I have to keep reminding them a lot of the kids are orphans because their parents died of HIV/AIDS, kids themselves will have HIV and several students came home from the first day at

the second school just in tears and sort of scuttled away to their rooms. So it's kind of trying to find this balance between – you're sharing a house so you're constantly there so they need their own space to cry and write and think and have their own reflection time and then at breakfast we talk. (Anna)

Whilst the STMP facilitator-educators play a key role in disseminating cultural knowledge, students involved in these programmes are involved in a range of activities through which they directly interact with local people and cultures and it is through these interactions that much cultural learning takes place. For example, Susie explained that in Vietnam she works with specific universities and during their visit the Australian students are involved in 'one-on-one' with the Vietnamese students. She views this cultural exchange between host and guest students as the 'best aspect' of this STMP as the Australian students have to 'deposit themselves within that context'. Similarly, education students travelling to Africa who were exposed to different teaching systems are described by the facilitator as having a transformed understanding of global differences for opportunities:

> Because every single one of them ... are able to reflect upon the fact that the education that they see in these schools that we go visit ... They are all of them able to go: how does this happen and under what conditions and who are these children and who are the teachers and they learn the relationship between politics and education, location and education and what they can and can't do as teachers in terms of impacting children's lives (Anna)

A key goal of the government's funding for student mobility schemes is building global awareness and cultural competencies. Whilst skill testing and networking are relatively tangible consequences of STMPs, just how a student might become cross-culturally competent is more problematic. One interviewee expressed such competency in the following way:

> It's asking the student to reflect on what they think about people from other cultures, what they think about themselves, what are their expectations. So expectation management is actually part of it as well ... and I suppose the ideal is to get a student to the point where, when they go abroad, their eyes are open to experience everything that's there and to reflect back on it ... So a culturally competent student would just be someone who has the ability to do that double loop – reflect, and then reflect again, and then really critically reflect afterwards. So identify – What did I see? What did it mean? And how has that actually changed or not changed my perceptions and behaviours about that experience? And I guess it's something we could all be doing in everything we do. But specific to the overseas experience it's one of those development concepts that's not taught in academic classrooms, it's not part of it. But if we want our students to be going abroad we need to help them understand what that's going to mean. (Linda)

The suggestion here is that by travelling overseas, students can develop critical insights that would otherwise be impossible in the classroom. The implication is also that students need educators to assist them to evaluate their experiences. She went on to explain:

> So we're doing this because we want our students to be global citizens, but they're studying to get jobs at an increasingly competitive workplace. It's great for us to say do this and it will improve your employment outcomes but does it really? How are we helping the students articulate the value of their overseas experience? That's where I think there's a really big gap. So before we dramatically increase the number of short term projects facilitated by faculty, how are we helping those [students] that have already done it [to] articulate what it meant to them? (Linda)

Linda further stated that an intrinsic issue for students who have participated in a STMP is their ability to communicate the global identity created by the experience and she emphasises that the experience is not necessarily about travel and mobility but, rather, how the student views the impact of the STMP on themselves, their global citizenry, and their employability:

> A student could say 'Yeah I interned at the UN' ... But does that make that student a better candidate [for a job] than the guy that flipped burgers at Maccas? Probably not [because] the two experiences are not comparable. But the skills that they've developed from those experiences may be comparable. It's up to the student to articulate [the experiences] to ... demonstrate [their] resilience ... to [give examples] of a time when you faced a challenging environment and what you did and what steps you took to overcome that challenge or to improve your performance or whatever. Both of those students could talk very differently but the skills we're looking to illicit or for them to demonstrate are the same. I think for the students that go abroad sometimes it's – I went overseas full stop.

In an effort to encourage students to think about their place in the world and to communicate the newly global identity, the interviewees discussed various assessment techniques that they used. One facilitator-educator explained how he uses reflective journals to get students to think about the purpose of their trip:

> They all apply the experiences they've had to their future career and part of the way I get them to do that is they have to keep a journal ... A reflective journal [submitted] at the end. One of the marking criteria is to examine how this experience might contribute to your future career so they're actually sort of ... [forced to think it through]. (Trevor)

Another facilitator-educator explained that students are expected to present seminars on their return to share their learnings. This process:

> Teaches them to articulate their experience at that process and it engages the next group of students who then want to go on the trip. (Melanie)

The evidence presented above points to the importance of the facilitator-educator as mediator of the STMP experiences of students which have as their purpose graduate attributes for connected and socially responsible global citizens. Research on the student experience is therefore missing in this study, but there is plenty of literature that focuses on student outcomes (e.g. Stone & Petrick, 2013; Tarrant et al., 2011; Tarrant & Lyons, 2012; Tarrant, Lyons, et al., 2014; Tarrant, Rubin, & Stoner, 2014). The findings reported here go some way to unpacking the complexity of nurturing global citizens through educational travel.

Discussion and conclusion

This study has explored the mediation of student global citizenry from the perspectives of a group of university staff who are facilitator-educators of short-term outbound mobility programmes. Whilst this research cannot be generalised as representative of widespread opinions, and further research in this area is needed, we found that the educators we interviewed all recognised that the programmes they facilitated enabled students to take a step closer to becoming global citizens. However, in most scholarly definitions, a global citizen is understood as a person who regards themselves as a citizen of the world rather than any particular nation-state (Lyons et al., 2012). The participants in this study were not able to clearly articulate what a global citizen was. However, they all recognised that facilitating global awareness and self-awareness was fundamental to it, and that one of their principal roles was to nurture this learning outcome.

Despite the diversity of disciplinary backgrounds and locations of the global mobility programmes they facilitated, all staff interviewed in this study emphasised the importance of exposing Australian students to global experiences that made them more globally aware and that challenged their own world view. Moreover, the participants in this study placed a great deal of emphasis on the processes of reflection that enabled students to make sense of the experiences they were having. Such an approach has long-been the focus of experiential learning principles.

Schön (1987), for instance, proposed two types of experiential learning reflection: 'reflection-in-action' and 'reflection-on-action'. The first involves reflexively thinking on one's feet and directly engaging with situations as they arise to generate a new understanding of phenomena (Schön, 1987). The second, reflecting 'on-action', takes place after a situation has been encountered (Schön, 1987). This act of reflection is a form of self-evaluation relating to how one acted as they did, what happened, and why. According to Schön (1987), this reflective practice process provides the mechanism by which a practitioner can assimilate new knowledge into their existing professional practice and, likewise, it can lead to a modification of future practice.

In the case of the facilitator-educators in this study, the reflectivity on global citizenry that they encouraged largely took place informally in conversations with students during the overseas experience. Whilst more traditional reflection-on-action activities, such as reflective writing in journals and presentations, were encouraged by the facilitator-educators, the informal teaching that took place through, for example, dinner meetings with students and in conversations whilst en-route to new locations, were an essential part of learning toolkit used by the facilitator-educators in this study to encourage students to reflect in-action (Schön, 1987).

Of course, these informal activities essentially involved university staff engaging with students in ways that would not be typical on campus. The facilitation of reflecting in-action overlaps with what might be described as pastoral care activities. It is clear from the interviews examined in part here, that students call on facilitators to provide more than just educational insights and experiences, as oftentimes students find themselves in need of help to cope with such things as culture shock, illness, homesickness, and so on. The expectation that facilitator-educators are responsible for pastoral care was raised by study participants as a significant challenge of the facilitator-educator role, particularly as these needs sit outside the normal realms of the teacher-student relationship in a classroom setting. Therefore, it is essential that further research seeks to understand the demands on staff who facilitate overseas student mobility experiences. The emotional labour that is involved may not be something that academic staff are not equipped to provide. Arguably, if facilitator-educators are unable to guide students through the more personal challenges of their mobility programme, there may be real implications for student learning outcomes that jeopardise the educational goals of STMPs. Therefore, further research that explores the pastoral care role of university staff more directly would be valuable. Such research would move beyond understanding the mediator role of facilitator-educators, to examining the leadership role concerned with managing the social and emotional well-being of students in the nurturing of global citizens.

Disclosure statement

No potential conflict of interest was reported by the authors.

Funding

This work was supported by a Faculty of Business and Law Research Project Grant.

ORCID

Tamara Young http://orcid.org/0000-0001-8132-4194
Joanne Hanley http://orcid.org/0000-0002-0987-5490

References

Australian Government. (2016). *National strategy for international education 2025*. Retrieved from https://nsie.education.gov.au/sites/nsie/files/docs/national_strategy_for_international_education_2025.pdf

Chieffo, L., & Griffiths, L. (2004). Large-scale assessment of student attitudes after a short-term study abroad program. *Frontiers: The Interdisciplinary Journal of Study Abroad, 10*, 165–177.

Cohen, E. (1985). The tourist guide: The origins, structure and dynamics of a role. *Annals of Tourism Research, 12*(1), 5–29.

Cohen, E. (2004). The tourist guide: The origins, structure and dynamics of a role. In E. Cohen (Ed.), *Contemporary tourism: Diversity and change* (pp. 159–178). Oxford: Elsevier.

du Cros, H., & McKercher, B. (2015). *Cultural tourism* (2nd ed.). Oxon: Routledge.

Crossman, J. E., & Clarke, M. (2010). International experience and graduate employability: Stakeholder perceptions on the connection. *Higher Education, 59*(5), 599–613.

Dann, G. (1996). *The language of tourism: A sociolinguistic perspective*. Wallingford, CT: CAB International.

Department of Education and Training. (2016). *Endeavour mobility grants*. Retrieved from https://internationaleducation.gov.au/Endeavour20program/studentmobility/Pages/International20Student20Mobility20Programs.aspx

Department of Foreign Affairs and Trade. (2016). *New Colombo plan*. Retrieved from http://dfat.gov.au/people-to-people/new-colombo-plan/pages/new-colombo-plan.aspx

Desforges, L. (2000). Traveling the world: Identity and travel biography. *Annals of Tourism Research, 27*(4), 926–945.

Dolby, N. (2007). Reflections on nation: American undergraduates and education abroad. *Journal of Studies in International Education, 11*(2), 141–156.

Giddens, A. (1991). *Modernity and self-identity: Self and society in the late modern age*. Stanford, CA: Stanford University Press.

Harrison, J. K. (2006). The relationship between international study tour effects and the personality variables of self-monitoring and core self-evaluations. *Frontiers: The Interdisciplinary Journal of Study Abroad, 13*, 1–22.

Harrison, L., & Potts, D. (2016). *Learning abroad at Australian universities: The current environment*. Melbourne: International Education Association of Australia.

Hendershot, K., & Sparandio, J. (2009). Study abroad and development of global citizen identity and cosmopolitan ideals in undergraduates. *Current Issues in Comparative Education, 12*(1), 45–55.

Hogg, M. A. (2006). Social identity theory. *Contemporary Social Psychological Theories, 13*, 111–1369.

Holland, J., & Thomson, R. (2009). Gaining perspective on choice and fate: Revisiting critical moments. *European Societies, 11*(3), 451–469.

Hovland, K. (2009). Global learning: What is it? Who is responsible for it? *Peer Review, 11*(4), 4.

Jennings, G., & Weiler, B. (2006). Mediating meaning: Perspectives on brokering quality tourist experiences. In G. Jennings & N. Polovitz Nickerson (Eds.), *Quality Tourism Experiences* (pp. 57–78). Burlington, MA: Elsevier Butterworth-Heinemann.

Karshenas, M. (1996). A step towards other cultures to give a chance to peace. In M. Robinson, N. Evens, & P. Callaghan (Eds.), *Tourism and cultural change* (p. 130). Sunderland: British Educational Publishers.

Ketabi, M. (1996). The socio-economic, political and cultural impacts of tourism. In M. Robinson, N. Evans, & P. Callaghan (Eds.), *Tourism and cultural change* (p. 155). Sunderland: British Educational Publishers.

van 't Klooster, E., van Wijk, J., Go, F., & van Rekom, J. (2008). Educational travel: The overseas internship. *Annals of Tourism Research, 35*(3), 690–711.

Lincoln Commission – Commission on the Abraham Lincoln Study Abroad Fellowship Program. (2005). *Global competence and national needs*. Washington, DC: Lincoln Commission.

Lyons, K. D. (2010). Room to move? The challenges of career mobility for tourism education. *Journal of Hospitality and Tourism Education, 22*(2), 51–55.

Lyons, K. D., Hanley, J. E., Wearing, S., & Neil, J. (2012). Gap year volunteer tourism. Myths of global citizenship? *Annals of Tourism Research, 39*(1), 361–378.

Mason, P., Augustyn, M., & Seakhoa-King, A. (2010). Exploratory study in tourism: Designing an initial qualitative phase of sequenced, mixed methods research. *International Journal of Tourism Research, 12*, 432–448.

McKeown, J. S. (2009). *The first time effect: The impact of study abroad on college student intellectual development*. Albany, NY: State University of New York Press.

McLaughlin, J. S., & Johnson, D. K. (2006). Assessing the field course experiential learning model: Transforming collegiate short-term study abroad experiences into rich learning environments. *Frontiers: The Interdisciplinary Journal of Study Abroad, 13*, 65–85.

Noy, C. (2004). This trip really changed me: Backpackers' narratives of self-change. *Annals of Tourism Research, 31*(1), 78–102.

Olsen, A. (2013). *Outgoing international mobility of Australian university students, 2012*. Presentation at the Australian International Education Conference, Canberra, Australia.

O'Reilly, C. (2006). From drifter to gap year tourist: Mainstreaming backpacker travel. *Annals of Tourism Research, 33*(4), 998–1017.

Pitman, T., Broomhall, S., McEwan, J., & Majocha, E. (2010). Adult learning in educational tourism. *Australian Journal of Adult Learning, 50*(2), 219–238.

Reysen, S., & Katzarska-Miller, I. (2013). A model of global citizenship: Antecedents and outcomes. *International Journal of Psychology, 48*(5), 858–870.

Schön, D. (1987). *Educating the reflective practitioner*. San Francisco, CA: Jossey-Bass.

Stone, M. J., & Petrick, J. F. (2013). The educational benefits of travel experiences in a literature review. *Journal of Travel Research, 52*(6), 731–744.

Stoner, K. A., Tarrant, M., Stoner, L., Perry, L., Wearing, S., & Lyons, K. D. (2014). Global citizenship as a learning outcome of educational travel. *Journal of Teaching in Travel and Tourism: The Professional Journal of the International Society of Travel and Tourism Educators, 14*, 149–163.

Sutton, R. C., & Rubin, D. L. (2004). The GLOSSARI project: Initial findings from a system-wide research initiative on study abroad learning outcomes. *Frontiers: The Interdisciplinary Journal of Study Abroad, 10*, 65–82.

Tarrant, M. A., & Lyons, K. (2012). The effect of short-term educational travel programs on environmental citizenship. *Environmental Education Research, 18*, 403–416.

Tarrant, M. A., Rubin, D., & Stoner, L. (2014). The added value of study abroad: Fostering a global citizenry. *Journal of Studies in International Education, 18*(2), 141–161.

Tarrant, M. A., Stoner, L., Borrie, W. T., Kyle, G., Moore, R. L., & Moore, A. (2011). Educational travel and global citizenship. *Journal of Leisure Research, 43*(3), 403–426.

Tarrant, M., Lyons, K., Stoner, L., Kyle, G. T., Wearing, S., & Poudyal, N. (2014). Global citizenry, educational travel and sustainable tourism: Evidence from Australia and New Zealand. *Journal of Sustainable Tourism, 22*(3), 403–420.

Thomson, R., Bell, R., Holland, J., Henderson, S., McGrellis, S., & Sharpe, S. (2002). Critical moments: Choice, chance and opportunity in young people's narratives of transition. *Sociology, 36*(2), 335–354.

University of Newcastle. (2015). *New futures strategic plan 2015–2025*. Retrieved from https://www.newcastle.edu.au/about-uon/our-university/vision-and-strategic-direction/new-futures-strategic-plan-2016-2025

Wearing, S., Stevenson, D., & Young, T. (2010). *Tourist cultures: Identity, place and the traveller*. London: Sage.

Wearing, S., Tarrant, M., Schweinsberg, S., Lyons, K., & Stoner, K. (2015). Exploring the global in student assessment and feedback for sustainable tourism education. In G. Moscardo & P. Benckendorff (Eds.), *Education for sustainability in tourism a handbook of processes, resources, and strategies*. (pp. 101–115). New York, NY: Springer.

Williams, T. R. (2005). Exploring the impact of study abroad on students' intercultural communication skills: Adaptability and sensitivity. *Journal of Studies in International Education, 9*(4), 356–371.

Young, T. (2009). Framing experiences of Aboriginal Australia: Guidebooks as mediators in backpacker travel. *Tourism Analysis, 14*(2), 155–164.

Young, T., & Lyons, K. D. (2011). 'I travel because I want to learn': Backpackers and the conduits of cultural learning. *Lifelong Learning in Europe, 15*(3), 150–158.

Young, T., & Lyons, K. D. (2016, June 26–29). *Exploring the institutionalisation of an ethic of care in outbound mobility programs*. Tourism Education Futures Initiative (TEFI) 9, Thompson Rivers University, Kamloops.

Educating tourists for global citizenship: a microfinance tourism providers' perspective

Giang Thi Phi, Michelle Whitford, Dianne Dredge and Sacha Reid

ABSTRACT
Ethical tourism initiatives have increasingly been framed as tools to educate tourists about global citizenship (GC), yet it is unclear how these initiatives are conceptualised, planned and implemented by tourism providers. This paper focuses on a form of ethical tourism known as microfinance tourism (MFT). It critically explores MFT providers' perspectives on what constitutes the goals of educating tourists about GC and how MFT can be designed and implemented to achieve these goals. The study adopted a qualitative approach utilising in-depth interviews with 12 key informants from 6 MFT organisations in Tanzania, Mexico, Jordan and Vietnam. The results reveal that MFT providers rely on an experiential learning process to educate tourists. However, as part of this learning process, MFT initiatives are located on a continuum, constituting those initiatives designed to increase tourists' compassion and philanthropic actions (i.e. 'thin' GC) through to those initiatives seeking to build solidarity and global discussions between tourists in order to challenge the structures that perpetuate global injustice (i.e. 'thick' GC). These results highlight the diversity of tourism providers' perspectives pertaining to GC, the effect diversity has on the design of tourism initiatives and the resultant outcomes of GC education utilising ethical tourism.

Introduction

The concept of global citizenship (GC) has a long history, which can be traced back to Socrates' notion of 'citizen of the world' and the cosmopolitanism of Enlightenment philosophers (Nussbaum, 2002; Parekh, 2003). In the last few decades, the rise of globalisation and the sustainable development agenda has prompted renewed philosophical and practical interest in GC. Its contemporary meanings have advanced from the simple notion that all human beings belong to a shared moral community, to incorporate a call for citizens to build awareness of their own and others' roles within an interconnected global context; to take responsibility for their actions; and to take actions towards development outcomes that are socially, economically, environmentally and politically just and sustainable (Dobson, 2003, 2006; Lapayese, 2003; Shultz, 2007).

In line with this trend, many ethical tourism initiatives have been promoted as tools to educate tourists about global issues and to foster an awareness that their tourism activities can create a better world (Butcher, 2015; Palacios, 2010; Pritchard, Morgan, & Ateljevic, 2011; Wearing, 2001). However, claims of their effectiveness remain largely unsubstantiated due to a lack of empirical research (Lyons, Hanley, Wearing, & Neil, 2012). Empirical studies have tended to overlook tourism providers' perspectives and little is known regarding how diverse tourism providers conceptualise, design and implement ethical tourism initiatives that will achieve the goals of GC education (Phi, Dredge, & Whitford, 2013). Arguably, this research-practice gap not only generates problems for the development of existing and new ethical tourism initiatives, but it also contributes to heightening scepticism surrounding tourism's potential for fostering GC (see e.g. Lyons et al., 2012; Tiessen & Huish, 2014).

This paper aims to critically explore GC education from the tourism providers' perspective. Three main questions guided the present study: What are the tourism providers' perceptions of GC education and its goals? What are the underlying factors that underpin the educational process in ethical tourism? And how do these diverse perceptions translate into the design and implementation of ethical tourism initiatives? To address these questions, we utilised a qualitative case study of microfinance tourism (MFT), which is a relatively new form of ethical tourism specifically developed to address global poverty issues by increasing tourists' awareness and actions (Sweeney, 2007).

The research is based on the premise that tourism providers have significant influence in the planning and implementation of ethical tourism initiatives and therefore play a key role in determining the outcomes of GC education. By advancing our understanding of providers' perspectives, this paper contributes to debates on how tourism can be used to educate tourists of GC. It also assists ethical tourism planners and providers, especially MFT providers, to better understand the complexity of GC education. This information can then, in turn, facilitate the development and implementation of initiatives that more effectively utilise and promote the values of GC to tourists.

Tourism and GC education: the gap in providers' perspective

GC, tourism and education are increasingly linked for a number of reasons. These include growing tourism's contribution to globalisation and increased global mobility, which in turn increases exposure of tourists (generally from developed economies) to poorer populations; increased awareness of others and differences in health, welfare, education, economic conditions; and heightened awareness of the interconnectedness of social, environmental and political issues that transcend geographical borders (Balarin, 2011; Carter, 2013; Lyons et al., 2012; Munar, 2007). Early research on tourism and GC often focused on socio-cultural impacts of tourism manifested through host–guest relationships and made a number of positive claims regarding how tourism can help to enhance international and cross-cultural understanding between individuals and nations, and promote global harmony and peace (e.g. Ketabi, 1996; Levy & Hawkins, 2010; Matthews, A., 2008). Along with the advancement of globalisation and GC theories, there have also been recent calls to also explore tourists' political power (e.g. Cameron, 2014; Munar, 2007). For instance, Munar (2007, p. 111) argued that:

> A tourist without a political dimension, without rights and duties, can never be the ground for sustainability. It is an issue that is not only about conserving the environment or enhancing the local culture, but fighting for human dignity.

The transformational approach to tourism acknowledges that 'in both the North and South, there exist concentrations of wealth and power along with increasing poverty and exclusion' (Shultz, 2007, p. 255) and it embraces the notion that tourists are political actors. There is thus a growing global consensus over the need for citizens, regardless of their nationality, social status and level of privilege, to share responsibility for actively addressing global social justice and help improve the lives of the poor and marginalised (McGrew, 2000). Therefore tourists are no longer seen merely as global consumers but rather, they become caring global citizens that have a responsibility to address global concerns (Donyadide, 2010).

One of the most common ways for tourism to bring global concerns to tourists' attention is via 'ethical' tourism initiatives (e.g. volunteer tourism, justice tourism, MFT), where educating tourists of GC is often claimed to be an end goal. Advocates of these initiatives claim that a range of positive outcomes can be realised including increasing tourists' compassion and understanding towards disadvantaged and marginalised groups (Gartner, 2008; Scheyvens, 2002, 2012; Wearing, 2001); raising awareness of, and commitment to, combating existing unequal power relations (Devereux, 2008; Wrelton, 2006); and building networks that promote activism and new social movements (Higgins-Desbiolles, 2011; McGehee, 1999; McGehee & Santos, 2005).

There is, however, a growing stream of literature that critically questions the role of tourism in fostering GC (e.g. Bianchi & Stephenson, 2013; Butcher, 2015; Butcher & Smith, 2010; Lyons et al., 2012; Sin, 2009; Tiessen & Huish, 2014). In what Butcher (2003) terms 'the moralisation of tourism', tourism consumption is given a moral agenda, promoted as a form of social action and an important solution in addressing underdevelopment and poverty issues in the developing world. This moral agenda encourages tourists to take small actions through their lifestyle and consumption choices, but these directives rarely involve educative elements that awaken tourists to their political power and responsibilities (Butcher & Smith, 2015). Put simply, GC education in tourism is frequently divorced from power and politics, which are key factors in addressing the root causes of global issues such as poverty. Hence, far from the promise of creating responsible global citizens, ethical tourism activities often result in the simplification of development and the fostering of tourists' unrealistic expectations about the extent to which their lifestyle choices/actions can help to address developmental issues (Baptista, 2012; Guttentag, 2011; Hutnyk, 1996; Lyons et al., 2012; Raymond & Hall, 2008; Vodopivec & Jaffe, 2011).

This debate over tourism and GC heightens the need for more empirical research on the impacts of ethical tourism initiatives on tourists, the design and implementation of such initiatives and the capacity of such initiatives to effectively educate tourists *for* GC, as opposed to education *about* GC. This distinction is important and will be developed later in the paper, but for now the latter (education *about* GC) can be understood as a more superficial transfer of information about the consequences of poverty. The former, (education *for* GC), conveys a deeper action-oriented commitment, and a more self-reflexive engagement with poverty's root causes and how one can become part of a social movement to address the poverty issue.

The literature generally focuses on the latter, describing tourists' perspectives, investigating the change in their values and actions after participating in ethical tourism (e.g. Broad, 2003; Brown & Morrison, 2003; Campbell & Smith, 2006; McGehee & Santos, 2005; Sin, 2009; Wearing, 2001; Zahra & McIntosh, 2007). How ethical tourism helps to foster GC has also been explored through the tourists' lens, which often attributed GC education as not intentionally organised or planned, but rather as a by-product of tourists' direct (and in volunteer tourism's case, prolonged) experience in the destination (Coghlan & Gooch, 2011; Tiessen & Huish, 2014; Wrelton, 2006). To date, little attention has been given to the contributions made by ethical tourism providers to foster tourists' commitment to and understanding of GC, despite their significant role in planning and implementing ethical tourism initiatives. Consequently, little is known about how tourism providers perceive the need for, and the goals of GC education, or their perceptions of how ethical tourism initiatives should be designed and implemented to achieve the goals of GC. The next section provides rationale for the selection of MFT as a case study to explore ethical tourism providers' perspectives on educating tourists of GC

MFT case study background

The idea of MFT was first conceptualised by Trip Sweeney in his seminal article explaining how the combination of microfinance (i.e. the provision of microloans and other financial services to poor populations) and tourism could serve as a much needed ethical initiative designed to advance poverty alleviation and GC education (Sweeney, 2007). Since 2008, six MFT organisations and operational models have been set up in Mexico, Tanzania, Jordan and Vietnam (see Table 1).

Table 1. Microfinance tourism providers.

Name	Location
Fundación En Vía	Oaxaca, Mexico
Investours Mexico (now Human Connections)	Puerto Vallarta, Mexico
Investours Tanzania	Dar es Salaam, Tanzania
Microfinance and Community Development Institute (MACDI)	Hanoi, Vietnam
Bloom Microventures	Hanoi, Vietnam
Zikra Initiative	Amman, Jordan

In essence, MFT facilitates an opportunity for tourists to experience microfinance in action by paying a visit to a small group of poor local micro-entrepreneurs operating in their daily environment. The MFT process typically starts with a local micro-entrepreneur who wishes to receive a free or low-interest micro-loan for income-generating activities. A small group of tourists are then brought to the local area via an organised tour and are hosted at the potential borrowers' houses. Tourists can enjoy hands-on local cultural activities provided by local people, while also learning about poverty issues and about microfinance as a means for poverty alleviation (Bloom Microventures, 2016). The majority of MFT providers concurrently deliver both microfinance and tourism programmes (i.e. Bloom Microventures, MACDI, Fundación En Vía and Zikra Initiative), while a few choose to only run the tours and partner with local microfinance institutions for microfinance activities (i.e. Investours Tanzania and Mexico). Depending on the models of operations, profit derived from the microfinance tours will be used either to fully subsidise interest rates for the microfinance institutions' loans, or turned into a low-interest microloans which will be repaid and recycled to assist more poor families in the area to improve their lives (Fundación En Vía, 2016; Investours, 2016).

As a typical ethical tourism initiative that was designed to foster GC among tourists (Phi et al., 2013), MFT seeks to capture the ever-increasing number of responsible/ethical tourists who would like to assist with various local developmental needs (Butcher, 2003). Currently, these tourists often find themselves with limited options and usually engage by way of philanthropic monetary donations, or labour and skill transfer in volunteering projects. These forms of aid are often criticised as creating dependency and bring only limited short-term impacts on people living in poverty or impoverished communities (Polak, 2009; Taplin, 2014). MFT intentionally captures the profit generated from tourism activities in impoverished areas and channels it into financial services that directly and more sustainably support individuals and communities living there. Tourists are provided an opportunity to experience the value of local 'empowerment and progress', through 'observing and supporting the small, significant successes of people in poverty who are moving hopefully forward' (Sweeney, 2007, p. 1). By providing 'something new' in the market, MFT can also attract growing numbers of tourists looking for alternative or direct tourism experiences with local communities (Novelli, 2005), especially those who are growing cynical of the mooted benefits of volunteer tourism.

MFT providers mainly focus on delivering short day-trips and occasionally overnight microfinance tours. This differs remarkably to volunteer tourism, where GC tends to be fostered rather naturally via tourists' prolonged stay in the destination. Microfinance tours' short duration makes it easier for tourists to participate in MFT as part of their overall travel experience in a destination. However this time constraint heightens the need for educational elements to be clearly embedded as part of the tour designs to effectively educate tourists of development issues, and encourage their further engagement in poverty alleviation. Thus, tourism providers' roles in fostering GCs are very pronounced in MFT initiatives. Additionally, the wide geographical distribution of the identified MFT providers also helped to enrich the case and enhance the diversity of the findings.

Data collection and analysis

Data collection strategies sought to gather information about tourism providers' perspectives on GC education. Semi-structured in-depth interviews with 12 respondents from six MFT organisations across four countries were undertaken to explore the multiple viewpoints regarding the conceptualisation, planning and implementation of MFT to foster GC (see e.g. Jennings, 2010; Stake, 2010). This enabled those respondents who were experiencing the phenomena to discuss the issues from their perspective and to explore the issues in depth (Boyce & Neale, 2006). Information solicited from respondents focused on three main lines of questioning:

(1) What are MFT's goals in educating tourists of GC?
(2) What are the unique MFT characteristics/features which helps MFT to achieve these goals?
(3) Are there any implementation issues which obstruct MFT achieving the goals?

To conduct the interviews, communication technology such as Skype (internet video calls) or telephone was utilised as a low cost and practical solution to the wide geographical distribution of respondents in global MFT (Gray, 2009). Interviews took place between October 2014 and February 2016. The key informants were purposively selected as they were 'influential, prominent or well informed people in an organisation', who have the potential to provide the researcher with detailed information and insights regarding the organisation's activities and impacts (Marshall & Rossman, 2011, p. 113). Because all MFT organisations were very small in size (ranging between 2 and 5 staff), the 'key informants' were extended to include original founders of the MFT programme/organisation, current and previous director (s)/board of directors, key managers and other key staff who have been directly involved in the planning/ implementation of MFT. Table 2 reveals the

Table 2. List of informants.

Name	Position	Gender	Mode of interview
Interview 1	Business development intern	Male	Skype
Interview 2	Founder/Board of directors	Male	Phone
Interview 3	Founder/Director	Female	Face-to-face
Interview 4	Director	Female	Skype
Interview 5	Founder/ Board of directors	Female	Skype
Interview 6	Chief operating officer	Female	Skype
Interview 7	Business development officer	Female	Skype
Interview 8	Business development manager	Female	Face-to-face
Interview 9	Operations manager	Female	Face-to-face
Interview 10	Founder/Chief operating officer	Male	Skype
Interview 11	Director	Male	Phone
Interview 12	Founder/Director	Female	Skype

characteristics of the informants so that the information they provide can be contextualised in the analysis below.

Interview durations varied between 60 and 90 minutes. Interviews were audio recorded with consent from the respondents and later transcribed to ensure accuracy of data. Thematic analysis, a process widely used for data analysis in qualitative research (see e.g. Braun & Clarke, 2006; Bryman, 2004) was applied to analyse the interviews. Thematic analysis involves the identification and reporting of key patterns and themes that 'capture something important about the data in relation to the research question' (Braun & Clarke, 2006, p. 82). Thematic analysis thus enables the researcher to reduce the general dataset into rich stories and thick descriptions that are important features of a case study research design (Ryan & Bernard, 2003). The results of data analysis revealed MFT providers in relation to their perspectives of GC could be located on a continuum ranging from 'thin' to 'thick' GC education.

MFT providers' goals for GC education

The 'thin–thick' GC education continuum

In the above discussion, a distinction was identified between education *for* GC ('thick' GC education) and education *about* GC ('thin' GC education). The former conveys a deeper and more reflexive style of education wherein tourists question both how their own actions contribute to the poverty problem, as well as what might be their own political power to address poverty alleviation. The latter describes a thinner and more superficial education that does not incite deeper engagement with the issues or tourists' own political power.

A key goal that the majority of MFT respondents share is to foster tourists' compassion towards people living in poverty, which would lead them to take philanthropic actions to support poverty alleviation. Respondents 4 and 10, for example, stated:

> I feel that the most important thing with any non-profit initiative, any vision for the future for me, is compassion - people having compassion for each other; people coming from a place of understanding and mutual respect [Respondent 4]

> Like if they [tourists] go back to their countries, they may talk about it, they may be inspired to do something for the local charity there. In their communities they may be inspired to donate to another project; I think that's something that a project such as ours propels. [Respondent 10]

This perspective on GC education reflects what Cameron (2014, p. 31) termed 'thin' GC. The notion of 'thin' GC is closely linked to the ancient Greek ideology of cosmopolitanism, which argues that all human beings belong to a single community based on a shared moral responsibility. A 'thin' cosmopolitan citizenship would signal each individual's moral responsibilities to help other human beings, regardless of their nation-state (Nussbaum, 2002).

Leveraging an individual's general sense of compassion, 'thin' GC is often divorced from politics and political action. Instead, an individual's desire 'to help' can be facilitated by private organisations and NGOs (such as MFT organisations) through the provision of ethical and/or charitable products and services (Butcher & Smith, 2015). An individuals' consumption of these products generally fulfils their sense of responsibility 'to do something' (Standish, 2012). In the MFT case, many respondents also articulated a depoliticised perception of poverty alleviation in which poverty was regarded as 'the lack of assets and material resources' [Respondent 3] or the lack of basic needs/capabilities such as 'education, food, and access to healthcare' [Respondent 10]. In essence, these respondents have compassion for people less fortunate than themselves and usually help by providing relief through various aid outlets, without thinking about what constitutes the real root cause of poverty. Thus, the 'thin' GC perspective reflects the need to focus on developing MFT initiatives that will encourage tourists to provide direct assistance to poor populations through the purchase of microfinance tours and further donations in the areas they visit or in their hometowns.

'Thin' GC focuses on an externalised, instrumental view of moral responsibility, while 'thick' GC embraces the notion of politics, where each individual is seen as having a political and deeply relational view of responsibility to address global injustice and create a more equal ground for all human beings to define their own development (Andreotti, 2014). Dobson (2006) noted that a 'privileged' population who possess power to travel extensively, currently occupy the global space. Furthermore, while most of these individuals do not directly or intentionally create poverty, they are still held responsible for simply participating in, rather than challenging the systems that benefit themselves at the expense of others (e.g. by consuming goods/services produced by cheap labour) (Matthews, S., 2008; Pogge, 2002).

In the MFT case, respondent 7 also cited 'the domination of international corporations' and 'the aid/development orthodoxy' as key global structures that 'reinforce and perpetuate poverty conditions'. Additionally, respondent 5 asserted that 'the problem is that we're normally coming up with this wide solution of what we think people really need and want'. One of the key goals for GC education under 'thick' citizenship is to create a new form of a global political community that is more inclusive of diverse discourses and voices (Linklater, 2006). In addition, there is a need to communicate to the 'privileged' tourists that their primary obligations are not only to assist people living in poverty, but also to work for the reform of global institutions that perpetuate unjust practices (e.g. World Trade Organisation, International Monetary Fund) and harmful models of development (e.g. top-down approaches that ignore local needs) (Cameron, 2014). In the case of MFT, respondent 2 expressed similar goals for building a global political community and of fostering tourists' activism through MFT:

> I think what we're doing is we're building a constituency of people who care a lot about the cause of poverty alleviation, and are actively involved in the community that they visit but also more broadly we want them to become a sort of community of people around the world who engage with these issues on a much higher policy level [Respondent 2]

A more active learning process was emphasised by respondent 2, where tourists were assisted by tour guides and the MFT organisation to achieve a higher level of critical reflection and engage in broader dialogue that connects local-global issues and their actions/inactions both during and after the tours. Respondents 2's perspective therefore is aligned with the concept of 'thick cosmopolitanism' (Cameron, 2014, p. 30) or 'republican citizenship' (Butcher & Smith, 2015, p. 101), which calls for every tourist to participate in informed and sustained political actions aimed at ending the suffering of others.

Returning to the idea that 'thick' and 'thin' GC education exists on a continuum, the analysis revealed a very textured variegation of positions between these two contrasting ends. The respondents' perspectives revealed that education for/about GC exists and includes a number of elements shown in Figure 1 (e.g. political power, locus of power, action orientation, perception of

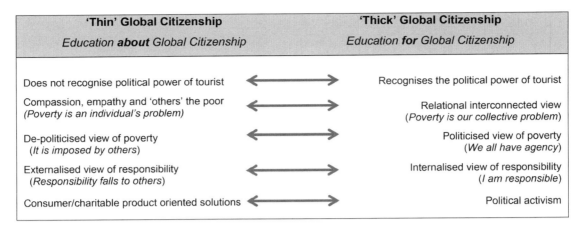

Figure 1. Thin and thick GC education.

responsibility, etc.). Moreover, while 'thin' and 'thick' descriptions help to anchor the end points on the continuum, it is important not to overgeneralise and treat 'thick' and 'thin' as homogenous positions occupying polar opposite positions on this continuum. Figure 1 depicts the various elements of 'thick' and 'thin' that were revealed in this study.

MFT providers' perceptions on the GC education

Advocates of ethical tourism initiatives often refer to experiential learning as the key pedagogy for GC education (e.g. Brigham, 2011; Raymond, 2008; Rennick & Desjardins, 2013; Tiessen & Huish, 2014). 'Experiential learning' is defined by Kolb (1984, p. 41) as 'the process whereby knowledge is created through the transformation of experience'. Kolb placed special focus on personal experience as the key that gives 'life, texture, and subjective personal meaning to abstract concepts' (Kolb, 1984, p. 21). He also identified four main stages of an experiential learning cycle where *concrete experiences* are seen as the crucial first step for the learning process. *Reflections* on these experiences then give rise to *abstract conceptualisation* (i.e. new insight or modifications of existing concepts), from which *implications for actions* emerge (Kolb, 1984).

In addition to Kolb, Schon (1983) and Mezirow (1991) have contributed important perspectives on the adult experiential learning process. For instance, Schon's (1983) 'learning-in-action' and Mezirow's (1991) transformative learning theories argue that for transformative changes to happen, the concrete experiences should have an element of a 'surprise' (Schon) or a 'dis-orientating dilemma' (Mezirow) in order to kick-start the process of challenging the attitudes, beliefs and experiences of learners. These authors also put more focus on the reflections (i.e. to think critically about the experiential activities) and dialogues (i.e. sharing one's experience/perspectives and listening to others) that take place during or after the experience. In other words, the interpretation and meaning attributed to the experience, rather than the experience itself is the key to enduring an individual's change in beliefs, attitudes and actions (Kirillova, 2015). MFT respondents in this study also revealed various key elements of this type of experiential learning.

Creating personal encounters (concrete experience)

Experiential learning emphasises the importance of providing the non-poor with opportunities to engage in concrete experiences with poorer populations. This type of concrete experience has been promoted in various ethical tourism initiatives (e.g. volunteer tourism, justice tourism, study abroad and occasionally slum tourism), where well-off tourists (often from developed countries) can visit and interact with the impoverished communities (Coghlan & Gooch, 2011; Scheyvens, 2012; Wearing, 2001). Similarly, MFT providers in this study identify 'creating personal encounters' as a key element for MFT. Respondent 12 viewed microfinance tours as 'helping to put a human face to what poverty looks like', while respondent 10 noted that MFT helps to facilitate a powerful 'first encounter' between people who are 'worlds apart'. He stated:

> The idea was that you've kind of established the first connection between two groups of people who are normally like worlds apart; like not only geographically but also in terms of their income, in terms of their education; in terms of their background. But you've created this first encounter; that's a very powerful thing that you can build upon. [Respondent 10]

Furthermore, respondents also stressed the 'personal' nature of these encounters as being important for GC education. Respondent 10 for instance, believed that 'What makes this very different is, let's call this "personal connections" that you are creating'. As tourism becomes increasingly commercialised, the community's commodified 'front stage' has long been the most common tourism setting (MacCannell, 1973). This is the space where tourists and the host merely observe each other from afar (e.g. see the tourist gaze; Urry & Larsen, 2011, and the host gaze; Moufakkir & Reisinger, 2013). Many researchers therefore have stressed the role of ethical tourism initiatives in providing tourists with interpersonal encounters that move beyond the simple 'gaze' (e.g. Chambers, 1983; Scheyvens, 2012; Wearing & Neil, 2000). Using examples of mass tourism and voyeuristic slum tours, MFT respondents also indicated that an impersonal tourism encounter would not qualify as a concrete experience for GC education purpose. In response, MFT providers provide tourists with the 'honour and privilege' [Respondent 12] to be 'invited into a local home or kitchen' [Respondent 9] and have personal conversations, 'going back and forth' with the hosts [Respondent 10].

Microfinance as new setting for tourism experience (surprise/disorienting dilemma)

The integration of microfinance in the tourism experience was seen by MFT respondents as providing an appropriate setting for challenging tourists' pre-conceptions regarding poverty and poverty alleviation. Respondents noted that many tourism initiatives tend to reinforce pity and stereotypes by exploiting 'overwhelmed images of poverty', where tourists or donors are taken to see 'a poor person with nothing', 'someone without slippers to walk on, or without food to eat' [Respondent 9]. Tourists who engage in ethical tourism initiatives are also frequently exposed to the sub-text of 'giving', 'saving' and 'helping' people living in poverty (Baptista, 2011; Sin, 2009).

In contrast, the traditional charitable act of one party 'giving, handing the money' [Respondent 5] or other forms of support (i.e. labour/material), to the other party is eliminated in MFT. Tourists are offered a quality tourism experience 'that's worth the price that they pay', in separation to the 'good cause' of poverty alleviation [Respondent 8]. Importantly, people living in poverty are not portrayed as 'the needy other' (Baptista, 2011, p. 663) in MFT. Through micro-loan investments, the tours help to showcase their strengths, commitment and efforts as microfinance clients, who work hard to improve their lives and to repay the capital they borrow. Thus the microfinance setting in MFT can be seen as a surprise or disorienting dilemma for tourists and provides the starting point for them to reflect on their assumptions about poverty and poverty alleviation. The comments of respondents 5 and 12 best summarise this idea:

> I think one very important difference that we were doing is to take away this mindset of people helping people ... you know, many times you can get a lot more from the locals than you can get from outside. So the help wasn't going directly from the tourists to the people, they're actually lending the money, so local people had the dignity part in it as well [Respondent 5]

> I think what our program does that is different from other programs that tourists may do is that they see that sort of strengths of the people, of the borrowers. They see that they have ideas, they see that they are hardworking, they see that they are committed, and this sort of agent of their own change [12]

Overall, MFT respondents' unanimously perceived microfinance tours as providing tourists with a concrete personal experience and disorienting dilemma (Mezirow, 1991) that triggers the GC education process. Importantly, it is this experiential learning component of the education process that appears to make a difference.

Tour guides' roles in assisting tourists' reflections and dialogues

Reflections and dialogues are integral elements of experiential learning. Several authors have argued that high level of critical reflection, along with active dialogue with diverse parties, is more likely to result in tourists building new insight and undergoing transformative changes (see e.g. Jacobson & Ruddy, 2004; Malinen, 2000; Mezirow, 1991; Percy, 2005). In this case study, MFT respondents emphasised the tour guides' roles in assisting tourists with this process of reflection and dialogue-building. However, the level of reflection and dialogue that tour guides are required to foster differed between some MFT respondents who facilitated 'thin' GC compared to other MFT respondents who facilitated 'thick' GC goals.

Many MFT respondents expected tourists who had dialogue with people living in poverty and also had a direct experience of their daily environment would allow those tourists to self-reflect on their previous assumptions regarding people living in poverty and develop more compassion for them. Respondent 4 commented:

> I think the idea is that if you are able to meet people from other cultures, experience other ways of life, understand some of the struggles that other people face, then it

fosters a greater awareness and cultural understanding that leads to more compassion. [Respondent 4]

From this perspective then, a major role for MFT tour guides is to 'facilitate the conversations between tourists and borrowers' [Respondent 6]. In particular, they need to 'set ground rules' for tourists' behaviours, perform 'cross-cultural translation – not only translating words but translating certain ideas or having cultural awareness of both sides' [Respondent 10], and 'make sure that the borrowers are comfortable and are able to tell their stories' [Respondent 12]. Besides assisting tourists to build dialogues with people living in poverty, MFT tour guides are seen as playing a more passive role, 'just let the people experience for themselves' [Respondent 10] and assist tourists' self-reflections by 'answering any questions they may have' [Respondent 9].

Heron (2011) argued that simply allowing tourists to self-reflect through interactions with people living in poverty is insufficient to achieve the goal of 'thick' GC education. Rather, to foster a more critical reflection process and build broader insight, ethical tourism organisations should actively engage tourists in critical reflection and dialogue that promotes linkages of 'causal responsibility'. This line of self-reflection links their actions or non-actions with the conditions of the people they visited and who are living in poverty (see Dobson, 2006; Tiessen & Huish, 2014). The emphasis on causal responsibilities, duties and obligations are seen as much stronger motivators for 'thick' GC actions (i.e. political activism) compared to simply appealing to tourists' compassion and sense of humanity (Dobson, 2006; Linklater, 2006).

The perceptions of respondent 2 were grounded in thick 'GC' values, suggesting that MFT tour guides should play a more active role to 'really engage the people that are travelling in discussions', for example:

> Through the tour, in the introduction and especially in the final debrief, talk to them about the more challenging issues of, kind of an intersection between tourism and development and poverty. Ask them to grapple with some of the more difficult challenges that we, as an organization, are facing. Ask them 'do you think it's ethical for you to come in to this community and do X Y Z'? [Respondent 2]

By encouraging tourists to openly share and/or discuss their experiences and perspectives, MFT tour guides can assist tourists to learn from their peers, instigating a form of collaborative learning that is crucial to broaden one's insight and challenge their world views (see Coghlan & Gooch, 2011; Cranton, 1994). Concurrently, MFT tour guides can direct the dialogue towards the broader ethical issues of global poverty and development and this can assist each tourist to reach a higher level of critical reflection beyond the local context and the people they visit, towards reflection on the unequal power relations between the well-off tourists and local communities, as well as considering the deeper structural causes of poverty.

Respondent 2 also noted that educational processes that help to foster 'thick' GC requires 'connotations and mental energy' from the tour guides. In addition, he also highlighted the need to 'better educate our tour guides in ethical issues'. For MFT tour guides to become effective facilitators of ethical dialogue between tourists and assist them to critically reflect beyond the context of microfinance tours, the MFT providers should 'have even more sophisticated training' which enables the tour guides to not simply 'bring up' but also 'know how to deal with and discuss these issues' [Respondent 2].

Overall however, MFT respondents reported that the tourists' reflections and dialogues facilitated by MFT tour guides generally lead to positive GC education outcomes such as a change in awareness and intentions, which can occur before concrete actions are taken in the next stage of the experiential learning cycle. For instance, Respondent 10 recalled notable changes in tourists' awareness regarding poverty at the end of the tour:

> We have cases where you could really see that people who came in with a certain mind-set in the morning, left with a different mind-set in the evening. I think that's very powerful and that's something which makes MFT so different from microfinance, and also which makes microfinance tourism different from classic tourism. Just this kind of combination of raising awareness. [Respondent 10]

Respondent 2 also reported tourists' positive intentions following the tour were a direct result of new insight gained from the experiential learning experiences:

> So first of all, I remember them articulating how powerful the experience was, but what showed us more was the fact that they kept reaching out and asking how they could do more. [Respondent 2]

MFT respondents' comments are supported by Nance's (2013) survey of 88 microfinance tourists in Mexico. This research reported tourists' strong intention to continue poverty alleviation actions after participating in MFT. Importantly however, Ballantyne, Packer, and Sutherland (2011) argued that it would be unlikely for the cycle of experiential learning to be completed during short tours and therefore, tourism organisations should provide tourists with post-tour support that encourages the ongoing development of reflection, dialogue and insight. Furthermore, tourists are more likely to act if they are given clear suggestions and opportunities on suitable courses of action (Ballantyne et al., 2011; Coghlan & Gooch, 2011). This earlier finding was also found in MFT respondents. For instance, Respondent 1 stated: 'I think it will be more effective if at the end of the day there is "a call for action" following the tour'.

MFT providers' post-tour support (reflections, dialogues and actions)

While most MFT respondents emphasised the importance of post-tour support in extending tourists' processes of reflection, dialogue and action, not surprisingly, respondents who perceived 'thin' GC goals identified the need for MFT organisations to provide different forms of post-tour support to those who perceived 'thick' GC goals.

The majority of MFT respondents articulated two main forms of support for tourists post-visit: (1) providing follow-up updates of borrower conditions; and (2) donation opportunities. First, MFT respondents showed attempts to develop solidarity between tourists and the local people by 'sending guests once every few months a few photos or short video clips about the borrowers' [Respondent 9]. These updates also served to remind tourists of their microfinance tour experience, in turn fostering further personal reflections and dialogues/sharing of the experience with others. Second, tourists were given options to continue their involvement with local poverty alleviation through online donations via the MFT organisations' websites. This is in line with Nance's (2013) suggestions that MFT organisations could increase the financial contribution per tourist by offering them immediate opportunities to donate and/or 'sponsor' other local micro-entrepreneurs. Both forms of post-tour support focus on increasing tourists' compassion and philanthropic actions to directly support people living in poverty. These are considered to be goals of 'thin' GC education.

In addition to the above post-tour support, respondent 2 also advocated for the development of an 'online global platform' which connects 'individuals and organisations who care about poverty alleviation and MFT'. The provision of an online global platform as part of post-tour support emphasises a MFT organisation's attempt to build solidarity among the tourists themselves, as well as between tourists and the broader civil society (e.g. local communities, NGOs, governments at different levels, tourism and microfinance organisations) to collectively take action to address the root causes of global poverty. Through this platform, tourists are encouraged to contribute their experience and ideas to a 'global discussion around microfinance, poverty, sustainable tourism etc' as 'informed participants' – 'people in countries around the world who have seen and experienced not just poverty but, more importantly, techniques in poverty alleviation' [Respondent 2]. Arguably then, the provision of a MFT online platform as post-tour support not only allows tourists to collectively take action to improve critically reflexive discourses on global poverty, it could also 'lead to improved policy' [Respondent 2]. Post-tour support could also serve to maintain the flow of critical reflection, dialogue and increased insight regarding poverty causes and poverty alleviation that were fostered during the tour. This form of post-tour support is therefore instrumental in achieving MFT's 'thick' GC goals of creating a global community that can 'act as a constituency for promoting actions against poverty in a myriad of ways' [Respondent 2].

Ironically, for Respondent 2's visions of building a global community for MFT and a global discussion surrounding poverty issues to be realised, MFT needs to grow and expand to become a global phenomenon, yet only one MFT organisation is currently pursuing this goal. MFT respondents' heavy focus on providing a MFT experience that fosters 'thin' GC is unsurprising, given that tourism is still predominantly a global industry which seeks to satisfy tourists' demand. Respondent 4 acknowledged this broader tourism context and stated that 'It's a tour, so it's meant to be an enjoyable experience, there's only so much information you can provide'. Respondent 9 noted that a 'more gentle version' would allow microfinance tours to appeal to 'a wider audience', including 'those who are not self-identified as being socially conscious or being responsible' [Respondent 9]. This is important for MFT organisations' financial sustainability and survival as their key revenue comes from tourism activities to support the whole operation (including the microfinance activities that directly support local people living in poverty).

Arguably however, without the integration of interventions to foster a higher level of critical reflection and broader dialogue from tourists (at least during the tour), MFT runs the very high risk of sending out simplistic/reductive messages about poverty and development. While microfinance is promoted as a key solution to poverty in MFT, microfinance itself represents a neoliberal approach to poverty alleviation, underlined by the assumption that poverty can be resolved simply by integrating people living in poverty into the market (Dini & Lippit, 2009). This use of market logic conveniently draws attention away from deeper causes of poverty, with a focus on calling for more donations of capital into microfinance sector to realise poor people's market potential (Harrison, 2008). Tourists' consumption of microfinance tours or their following philanthropic actions therefore would still mainly serve to make them 'feel good about themselves' or 'feel like they're making a difference' [Respondent 6], whilst not helping them in anyway to develop deep insight of, or take actions to, address poverty's root causes (e.g. the system of privileges that allow them to retain their 'superior' or 'luckier' positions to the people they visit – Sin, 2009).

Conclusion

This paper has explored the perspectives of ethical tourism (e.g. MFT) providers' in relation to GC education. The case study of MFT has demonstrated that these ethical tourism providers possess very diverse perspectives regarding what constitutes GC and the goals of GC education. These diverse perspectives heighten the complexity surrounding ethical tourism providers' efforts to educate tourists about GC. Different providers tend to conceptualise and implement different approaches to the education process which ultimately affect the outcomes of ethical tourism education.

Overall, MFT providers within this study relied on an extended experiential learning process to educate tourists of GC that extended beyond the duration of the microfinance tour. MFT respondents generally agreed that the integration of microfinance into the tourism experience provided a unique context for personal encounters between tourists and people living in poverty. Microfinance settings also provide the necessary surprise/disorienting dilemma that triggers the tourists' process of questioning previous assumptions regarding poverty and poverty alleviation. However, MFT respondents differ in terms of the perceived types of interventions that the MFT organisations should carry out to assist tourists in their learning process. It has been shown that these differences are rooted in the respondents' diverse perceptions regarding what constitutes GC and associated goals.

The results of this research highlight the current state of the ethical tourism market in relation to GC and corroborate the view that it is dominated by programmes and initiatives characterised by 'thin' GC (Tiessen & Huish, 2014). Framed by 'thin' GC goals, the majority of respondents conceptualised and implemented a passive learning process which allows tourists to self-reflect through observing and having conversations with people living in poverty. The results suggest that on one hand, ethical tourism may focus on tourists' role as active agents of social change. Yet on the other hand, tourists are often provided with depoliticised, fragmented and simplistic information of global issues, which result in actions that perpetuate voluntaristic/philanthropic approaches and unrealistic expectations regarding contributions to address global concerns.

Therefore, it is important for all key stakeholders involved in GC education through tourism (e.g. tourism providers, tour guides, tourists, funding organisations such as universities and development agencies, local communities, and governments) to be informed of the different perspectives regarding GC and make conscious decisions on the goals and practices of fostering GC. Ideally, GC education for these stakeholders will push them to the 'thick' end of the continuum. However, even if ethical tourism providers choose to promote 'thin' GC for practical reasons, they should be aware of its limitations and make efforts to integrate certain elements that assist tourists to think more critically and engage in broader dialogue, without negatively affecting the overall tourism experience. Furthermore, as the experiential learning process was shown to extend beyond the duration of the tour, ethical tourism providers should consider providing tourists with a wider range of opportunities to take actions after the tour that go beyond economic/labour transfer (e.g. participation in social/political movements). Avenues should also be provided for tourists to further reflect and engage in discussions regarding the issues that were brought to their attention during the tour, such as the development of online forums or networks.

The exploratory nature of this study has laid the foundation for future research opportunities. First, the study highlighted the importance of appropriate settings and personal encounters which trigger the experiential learning process via tourism. Future research should explore the range of tourism context/conditions that would be most conducive to tourists' transformative changes in attitudes, beliefs and actions. Second, the tour guides were shown to play a major role in the educational process. Though the tourism literature has started to pay attention to the power of tour guides in directing the host-guest exchange (e.g. Cheong & Miller, 2000; Mowforth & Munt, 2003; Wrelton, 2006), empirical research is still needed to explore more the extent to which tour guides' attitudes and practices affect outcomes of GC education, as well as the challenges they face during the educational process.

Acknowledgement

This research is supported by an Australian Government Research Training Program (RTP) Scholarship.

Disclosure statement

No potential conflict of interest was reported by the authors.

ORCID

Giang Thi Phi http://orcid.org/0000-0002-2359-2833

References

Andreotti, V. O. (2014). Soft versus critical global citizenship education. In S. McCloskey (Ed.), *Development education in policy and practice* (pp. 21–31). London: Palgrave Macmillan.

Balarin, M. (2011). Global citizenship and marginalisation: Contributions towards a political economy of global citizenship. *Globalisation, Societies and Education*, *9*(3–4), 355–366.

Ballantyne, R., Packer, J., & Sutherland, L. A. (2011). Visitors' memories of wildlife tourism: Implications for the design of powerful interpretive experiences. *Tourism Management*, *32*(4), 770–779.

Baptista, J. (2011). The tourists of developmentourism – representations 'from below'. *Current Issues in Tourism*, *14*(7), 651–667.

Baptista, J. (2012). Tourism of poverty: The value of being poor in the nongovernmental order. In F. Frenzel, K. Koens, & M. Steinbrink (Eds.), *Slum tourism: Poverty, power and ethics* (pp. 125–143). London: Routledge.

Bianchi, R. V., & Stephenson, M. L. (2013). Deciphering tourism and citizenship in a globalized world. *Tourism Management*, *39*, 10–20.

Bloom Microventures. (2016). *How it works*. Retrieved from http://bloommv.org/microfinance/how-it-works/

Boyce, C., & Neale, P. (2006). *Conducting in-depth interviews: A guide for designing and conducting in-depth interviews for evaluation input*. Watertown, NY: Pathfinder International.

Braun, V., & Clarke, V. (2006). Using thematic analysis in psychology. *Qualitative Research in Psychology*, *3*(2), 77–101.

Brigham, M. (2011). Creating a global citizen and assessing outcomes. *Journal of Global Citizenship & Equity Education*, *1*(1), 15–43.

Broad, S. (2003). Living the Thai life – a case study of volunteer tourism at the Gibbon Rehabilitation Project, Thailand. *Tourism Recreation Research*, *28*(3), 63–72.

Brown, S., & Morrison, A. M. (2003). Expanding volunteer vacation participation an exploratory study on the mini-mission concept. *Tourism Recreation Research*, *28*(3), 73–82.

Bryman, A. (2004). *Social research methods* (2nd ed.). New York, NY: Oxford University Press.

Butcher, J. (2003). *The moralisation of tourism: Sun, sand … and saving the world?* London: Routledge.

Butcher, J. (2015). Ethical tourism and development: The personal and the political. *Tourism Recreation Research*, *40*(1), 71–80.

Butcher, J., & Smith, P. (2010). 'Making a difference': Volunteer tourism and development. *Tourism Recreation Research*, *35*(1), 27–36.

Butcher, J., & Smith, P. (2015). *Volunteer tourism: The lifestyle politics of international development*. New York, NY: Routledge.

Cameron, J. (2014). Grounding experiential learning in 'thick' conceptions of global citizenship. In R. Tiessen & R. Huish (Eds.), *Globetrotting or global citizenship* (pp. 21–42). Toronto: University of Toronto Press.

Campbell, L. M., & Smith, C. (2006). What makes them pay? Values of volunteer tourists working for sea turtle conservation. *Environmental Management*, *38*(1), 84–98.

Carter, A. (2013). *The political theory of global citizenship*. New York, NY: Routledge.

Chambers, R. (1983). *Rural development: Putting the last first*. New York, NY: Routledge.

Cheong, S.-M., & Miller, M. L. (2000). Power and tourism: A Foucauldian observation. *Annals of Tourism Research*, *27*(2), 371–390.

Coghlan, A., & Gooch, M. (2011). Applying a transformative learning framework to volunteer tourism. *Journal of Sustainable Tourism*, *19*(6), 713–728.

Cranton, P. (1994). *Understanding and promoting transformative learning: A guide for educators of adults*. San Francisco, CA: Jossey-Bass.

Devereux, P. (2008). International volunteering for development and sustainability: Outdated paternalism or a radical response to globalisation? *Development in Practice*, *18*(3), 357–370.

Dini, A., & Lippit, V. (2009). *Poverty, from orthodox to heterodox approaches: A methodological comparison survey (Working Paper)*. Riverside: University of California.

Dobson, A. (2003). *Citizenship and the environment*. Oxford: Oxford University Press.

Dobson, A. (2006). Thick cosmopolitanism. *Political Studies*, *54*(1), 165–184.

Donyadide, A. (2010). Ethics in tourism. *European Journal of Social Sciences*, *17*(3), 426–433.

Fundación En Vía. (2016). *Microfinance program*. Retrieved from http://www.envia.org/microfinance_program.html

Gartner, C. M. (2008). Tourism, development, and poverty reduction: A case study from Nkhata Bay, Malawi (*Master's thesis*). University of Waterloo.

Gray, D. (2009). *Doing research in the real world*. London: Sage.

Guttentag, D. (2011). Volunteer tourism: As good as it seems? *Tourism Recreation Research*, *36*(1), 69–74.

Harrison, D. (2008). Pro-poor tourism: A critique. *Third World Quarterly*, *29*(5), 851–868.

Heron, B. (2011). Challenging indifference to extreme poverty: Considering southern perspectives on global citizenship and change. *Ethics and Economics*, *8*(1), 110–119.

Higgins-Desbiolles, F. (2011). *Resisting the hegemony of the market: Reclaiming the social capacities of tourism*. Bristol: Channel View.

Hutnyk, J. (1996). *The rumour of Calcutta: Tourism, charity and the poverty of representation*. London: Zed Books.

Investours. (2016). *Investours – our story*. Retrieved from http://investours.org/about-investours/our-story/

Jacobson, M., & Ruddy, M. (2004). *Open to outcome: A practical guide for facilitating & teaching experiential reflection*. Bethany, OK: Wood'N'Barnes.

Jennings, G. (2010). *Tourism research* (2nd ed.). Milton: John Wiley and Sons.

Ketabi, M. (1996). The socio-economic, political and cultural impacts of tourism. In M. Robinson, N. Evans, & P. Callaghan (Eds.), *Tourism and cultural change* (pp. 155–170). Sunderland: British Education.

Kirillova, K. A. (2015). Existential outcomes of tourism experience: The role of transformative environment. *Doctoral thesis*, Purdue University, West Lafayette, IN, US.

Kolb, D. (1984). *Experiential learning*. Englewood Cliffs, NJ: Prentice Hall.

Lapayese, Y. V. (2003). Toward a critical global citizenship education. *Comparative Education Review*, *47*(4), 493–501.

Levy, S. E., & Hawkins, D. E. (2010). Peace through tourism: Commerce based principles and practices. *Journal of Business Ethics*, *89*, 569–585.

Linklater, A. (2006). Cosmopolitanism. In A. Dobson & R. Eckersley (Eds.), *Political theory and the ecological challenge* (pp. 109–128). Cambridge: Cambridge University Press.

Lyons, K., Hanley, J., Wearing, S., & Neil, J. (2012). Gap year volunteer tourism: Myths of global citizenship? *Annals of Tourism Research*, *39*(1), 361–378.

MacCannell, D. (1973). Staged authenticity: Arrangements of social space in tourist settings. *American Journal of Sociology*, *79*(3), 589–603.

Malinen, A. (2000). *Towards the essence of adult experiential learning: A reading of the Theories of Knowles, Kolb, Mezirow, Revans and Schon*. Jyväskylä, Finland: Sophi Academic Press.

Marshall, C., & Rossman, G. B. (2011). *Designing qualitative research* (5th ed.). Thousand Oaks, CA: Sage.

Matthews, A. (2008). Negotiated selves: Exploring the impact of local-global interactions on young volunteer travellers. *Journeys of Discovery in Volunteer Tourism: International Case Study Perspectives*, 101–117.

Matthews, S. (2008). The role of the privileged in responding to poverty: Perspectives emerging from the post-development debate. *Third World Quarterly*, *29*(6), 1035–1049.

McGehee, N. G. (1999). Alternative tourism: A social movement perspective (*Doctoral thesis*). University of Newcastle.

McGehee, N. G., & Santos, C. A. (2005). Social change, discourse and volunteer tourism. *Annals of Tourism Research*, *32*(3), 760–779.

McGrew, A. (2000). Sustainable globalization? The global politics of development and exclusion in the new world order. In T. Allen & A. Thomas (Eds.), *Poverty and development into the 21st century* (pp. 345–352). Oxford: Oxford University Press.

Mezirow, J. (1991). *Transformative dimensions of adult learning*. San Francisco, CA: Jossey-Bass.

Moufakkir, O., & Reisinger, Y. (2013). *The host gaze in global tourism*. Boston, MA: CABI.

Mowforth, M., & Munt, I. (2003). *Tourism and sustainability. New tourism in the third world*. London: Routledge.

Munar, A. M. (2007). Rethinking globalization theory in tourism. *Tourism Culture & Communication*, *7*(2), 99–115.

Nance, M. (2013). *Tourist intentions to continue poverty alleviation actions after participating in microfinance tourism*. Clemson, SC: Clemson University: TigerPrints.

Novelli, M. (2005). *Niche tourism. Contemporary issues, trends and cases*. Oxford: Butterworth-Heinemann.

Nussbaum, M. C. (2002). Education for citizenship in an era of global connection. *Studies in Philosophy and Education*, *21*(4-5), 289–303.

Palacios, C. M. (2010). Volunteer tourism, development and education in a postcolonial world: Conceiving global connections beyond aid. *Journal of Sustainable Tourism*, *18*(7), 861–878.

Parekh, B. (2003). Cosmopolitanism and global citizenship. *Review of International Studies*, *29*(1), 3–17.

Percy, R. (2005). The contribution of transformative learning theory to the practice of participatory research and extension: Theoretical reflections. *Agriculture and Human Values*, *22*(2), 127–136.

Phi, G., Dredge, D., & Whitford, M. (2013). *Fostering global citizenship through the microfinance-tourism nexus*. Paper presented at the 7th annual tourism education futures initiative conference 2013 (TEFI7): tourism education for global citizenship: Educating for lives of consequence Oxford, UK.

Pogge, T. W. (2002). *World poverty and human rights*. Cambridge: Polity Press.

Polak, P. (2009). *Out of poverty: What works when traditional approaches fail*. California, CA: Berrett-Koehler.

Pritchard, A., Morgan, N., & Ateljevic, I. (2011). Hopeful tourism: A new transformative perspective. *Annals of Tourism Research*, *38*(3), 941–963.

Raymond, E. (2008). 'Make a difference!': The role of sending organizations in volunteer tourism. In K. Lyons & S. Wearing (Eds.), *Journeys of discovery in volunteer tourism* (pp. 48–62). Oxfordshire: CABI.

Raymond, E. M., & Hall, C. M. (2008). The development of cross-cultural (mis) understanding through volunteer tourism. *Journal of Sustainable Tourism*, *16*(5), 530–543.

Rennick, J. B., & Desjardins, M. (2013). *The world is my classroom: International learning and Canadian higher education*. Toronto: University of Toronto Press.

Ryan, G. W., & Bernard, H. R. (2003). Techniques to identify themes. *Field Methods*, *15*(1), 85–109.

Scheyvens, R. (2002). *Tourism for development: Empowering communities*. London: Pearson Education.

Scheyvens, R. (2012). *Tourism and poverty*. New York, NY: Routledge.

Schon, D. A. (1983). *The reflective practitioner: How professionals think in action*. New York, NY: Basic Books.

Shultz, L. (2007). Educating for global citizenship: Conflicting agendas and understandings. *Alberta Journal of Educational Research*, *53*(3), 248–258.

Sin, H. L. (2009). Volunteer tourism – 'involve me and I will learn'? *Annals of Tourism Research*, *36*(3), 480–501.

Stake, R. E. (2010). *Qualitative research: Studying how things work*. New York, NY: Guilford Press.

Standish, A. (2012). *The false promise of global learning: Why education needs boundaries*. New York, NY: Continuum International.

Sweeney, T. (2007). *Why we need micro loans instead of slum tourism*. Retrieved from http://matadornetwork.com/bnt/why-we-need-micro-loans-instead-of-slum-tourism/

Taplin, J. E. (2014) Programme monitoring and evaluation practices of volunteer tourism organisations (*Doctoral thesis*). Southern Cross University, Lismore.

Tiessen, R., & Huish, R. (2014). *Globetrotting or global citizenship?: Perils and potential of international experiential learning*. Toronto: University Of Toronto Press.

Urry, J., & Larsen, J. (2011). *The tourist gaze 3.0*. London: Sage.

Vodopivec, B., & Jaffe, R. (2011). Save the world in a week: Volunteer tourism, development and difference. *European Journal of Development Research*, *23*(1), 111–128.

Wearing, S. (2001). *Volunteer tourism. Experiences that make a difference*. Oxon: CABI.

Wearing, S., & Neil, J. (2000). Refiguring self and identity through volunteer tourism. *Society and Leisure*, *23*(2), 389–419.

Wrelton, E. (2006). 'Reality tours' to Chiapas, Mexico: The role of justice tourism in Development (*Master's thesis in Development Studies*). Massey University, New Zealand.

Zahra, A., & McIntosh, A. (2007). Volunteer tourism: Evidence of cathartic tourist experiences. *Tourism Recreation Research*, *32*(1), 115–119.

The limits of cosmopolitanism: exchanges of knowledge in a Guatemalan volunteer programme

Rebecca L. Nelson

ABSTRACT
Drawing from 20 months of ethnographic fieldwork in the voluntourism programme of a women's weaving cooperative in Quetzaltenango, Guatemala, this article argues that cooperative leaders faced a dilemma in handling voluntourists. They sought to develop globally oriented market knowledge using voluntourists as consultants while protecting their own expertise from tourists as potential competitors in these foreign markets. Originating from the same communities as the cooperative's clients, voluntourists shared their understanding of tastes and practices in their home countries to help the organisation export weavings more effectively. In return, the cooperative leaders offered them exposure to Mayan customs and weaving traditions. However, these voluntouristic information exchanges also had limits: cooperative leaders were wary of sharing their traditional weaving practices with tourists. This mobilisation of their locally rooted, identity-defining knowledge made cooperative leaders anxious that tourists were establishing 'Mayan' weaving schools in their home countries. Voluntourists struggled less with the contradictions between belonging to a place and adapting to global citizenship than cooperative leaders because the voluntourists' knowledge had already been detached from its geographic roots and made transferrable.

Introduction

A key assumption in the literature on cosmopolitanism in tourism is that the circulation of tourists can make people more cosmopolitan by exposing them to new ways of life and modes of thought—not just the tourists themselves, but also those who play host to them (Clifford, 1992; Ferguson, 1999; Hannerz, 2006; Notar, 2008; Wardle, 2000; Werbner, 1999). Based on ethnographic research in the voluntourism programme of TelaMaya (a pseudonym), a women's weaving cooperative based in Quetzaltenango, Guatemala, this article argues that, indeed, in certain ways voluntourists and their cooperative hosts alike developed more globally oriented ways of thinking and acting through their daily work together. The cooperative leaders engaged with voluntourists primarily as consultants. Because they came from the same communities as the cooperative's clients, international voluntourists' main role in the cooperative was to share their knowledge of the tastes, expectations, and practices of their home countries with the weavers to help the organisation export its products more effectively. As the cooperative secretary, Paula, stated, 'It's very important to have the support of foreign people because they're people from other countries—they know their people, what it is they like when they come to Guatemala'. In return, the cooperative leaders offered to give volunteers insights into Mayan customs, practice speaking Spanish, and classes in traditional backstrap weaving. The Mayan weavers who run Tela-Maya were painfully aware that their prosperity was intimately linked to their ability to understand people from other places. At the same time, these interactions highlighted the limitations of what the organisation's members wanted to share, in the context of global trends towards defining cultural knowledge as a proprietary resource to be managed, protected, legislated, and exploited. They sought to partake in global flows of information and goods on their own terms, using voluntourists' advice to position their products as fair trade and rooted in ancient practices to satisfy tourists' expectations and command higher prices, without losing control of the asset that attracted tourists to them in exchange.

Cosmopolitanism has been variously described as a socio-cultural state of being, an outlook on the world, a political project for constructing transnational institutions and creating a space for plurality, and a set of skills that allow people to move across the globe (Szerszynski and Urry, 2006; Vertovec & Cohen, 2002). It is primarily in the last sense, as a form of competency that individuals can cultivate, that this article will investigate the utilitarian exchanges of knowledge within TelaMaya's voluntourism programme. A pair of central dilemmas were inherent in the exchanges of knowledge within the TelaMaya voluntourism programme: while voluntourists were drawn to the cooperative by their appreciation for Mayan customs, they sought to improve the cooperative by remaking it along Western lines. Also, the cooperative leaders sought to develop globally oriented market knowledge using voluntourists as consultants while protecting their own expertise from tourists as potential competitors in these foreign markets.

Entanglement with other world views: empowering or disempowering?

Both groups benefitted from these exchanges in cultivating their cosmopolitan orientations, perspectives, and skills. Through voluntourism, overeducated and underemployed youths from the US and Europe prepare for global citizenship by performing international humanitarianism (Vrasti, 2013). They demonstrate their capacity to operate in a foreign context and assume important-sounding responsibilities in a context in which their expertise and preparation may be questioned less than it would be in their home countries. Similarly, people who play host to regular flows of international tourists have an economic and political stake in viewing themselves as citizens of a larger world. They can use their exposure to foreign ways of life through personal interactions with tourists, as well as international media and other sources, to build their cultural capital (Salazar, 2010). Within this research, voluntourists were more inclined to describe themselves as global citizens than their hosts, because they have been encouraged to think of themselves in those terms, whereas Mayan weavers have not. However, the Mayan weavers were much more aware of current events in the US, for example, than US voluntourists were of Guatemala. Cosmopolitanism as an etic analytical term must be separated from people's emic identification with it. Some people may not be inclined to self-identify as cosmopolitan but nonetheless demonstrate cross-cultural competencies and orientations (Hannerz, 2006).

However, the openness to exchanges of cultural knowledge developed through these voluntouristic relationships had important limits. The cooperative leaders were wary of revealing their traditional weaving practices to tourists, in part because their dealings with tourists made them aware of the market value of Mayanness. This mobilisation of their locally rooted, identity-defining knowledge made the board of directors of the cooperative anxious that voluntourists could carry away their local knowledge to exploit it in other spaces and contexts. A concrete instance of this anxiety was their constant fear that foreigners were establishing imagined 'Mayan' weaving schools in their home countries and eroding their market. In this case, a cosmopolitan awareness of foreigners' valuation of indigenous knowledge helped influence a group of Guatemalan women to conceive of their patrimony as an economic and political resource and guard it against foreign incursions. This study closely examines the lived realities of disparate groups of people who are struggling to develop border-crossing skills, enacting and embodying ways of being that encompass more than one geographic locale, and at times challenging the tendencies of those who consider themselves cosmopolitan to gloss over entrenched power differentials in the pre-existing global terrain.

Scholars like Appiah (2006) have presented cosmopolitanism as a universal good, and those who seek to isolate themselves as parochial. However, cultural knowledge has increasingly become a site of contestation. The essentialised view of culture as a resource to be appropriated, revived, guarded, separated, and identified has become ubiquitous over the last few decades, taken up by tourists and indigenous people alike (Yúdice, 2003). This redefinition and reification of the concept of culture has provoked conflicts over cultural appropriation and the growing anxiety over cultural preservation over the past four decades (Brown, 2003). Thinking of cultural knowledge as a resource, and potentially a limited one, makes it a site of contestation.

Two important perspectives on cultural knowledge have emerged within the recent literature in the social sciences that roughly map onto universalising and particularising orientations. The first, cosmopolitanism apparently opens up possibilities and invites cultural comparison and potential change; the other, the view of cultural knowledge as a resource establishes boundaries around traditions. However, there are arguably points of convergence between these perspectives: viewing cosmopolitanism as a set of competencies fits well with the viewpoint that knowledge is a resource, and volunteers and organisations alike seek to become masters of it. Learning to 'mobilise' cultural knowledge as a resource (both in the sense of deploying it and translating it for ease of travel across various borders of understanding) can itself be a cosmopolitan form of knowledge. This article explores these points of convergence to understand how, depending on how the scholarly perspective is tilted, entanglement with other world views can be simultaneously emancipatory and disempowering.

Knowledge as a resource

The management of knowledge has been identified as the key to success in the ethical consumption market, in which consumers express their values through their purchasing decisions (Butcher, 2003; Hudson & Hudson, 2003; Princen, Maniates, & Conca, 2002; Taylor, 2005). TelaMaya voluntourists deeply understood the particular expectations of ethical consumers because they were participating in voluntourism, a form of ethical consumption, themselves. They sought to extract specific personal information from the cooperative members to make them legible to ethical consumers (Fair Trade buyers and voluntourists alike) and seek their aid. This dynamic makes Fair Trade marketing more intrusive for producers than 'typical' marketing where producers largely remain invisible (Bryant & Goodman, 2004; Lyon, 2006), just as scholars have noted that voluntourism can have a higher economic, socio-cultural, psychological, and environmental impact on local communities than mainstream tourism (Butler, 2004; Macleod, 2004), because of the more intimate and long-term relationships it creates between visitors and visited.

Fair Trade marketing has been critiqued for providing rich information about producers to consumers, while failing to inform producers about consumers; this one-way flow is problematic because Fair Trade becomes an exchange of information for money and resources, rather than a reciprocal relationship between producers and consumers (Lyon, 2006). However, this critique is complicated by the presence of voluntourists in cooperatives like TelaMaya, who provided information to the artisans about foreigners' tastes and interests. The volunteers in many ways personified the marketplace for their Mayan hosts—they were buyers (contributing directly to the cooperative by selling products to themselves and their friends and family) and stand-ins for the buyers. While scholars, volunteers, and weavers alike have tended to view the global marketplace as a faceless and mysterious system (Gibson-Graham, 2006), capitalist institutions are made up of individuals and sets of practices. For TelaMaya, volunteers *were* the market in a literal and figurative sense, standing in for imagined consumers and personifying a fickle and untrustworthy marketplace. Volunteers have become incorporated in the process of cultural production in TelaMaya. This process of representation created a push-pull dynamic between voluntourists and TelaMaya members in which the volunteers were driven to uncover and reveal deeper information about the Mayan weavers, who in turn sought to hold back intellectual possessions that were intimately linked to their identities.

In her classic work on reciprocal exchange in the Trobriand Islands, Weiner (1992) presented it as a field of struggle in which each group seeks to hold back certain elements associated with group identity from the exchange to reaffirm the boundaries between their groups. Aware that the other group is holding back, exchange participants strategically angle to gain access to that which is not being offered. Weiner (1992, p. 43) asserted that the differentiating power of inalienable possessions makes it impossible for a simple exchange to take place in which one good, idea, or service is given in return for another without participants being aware of the relationship between that possession and identity or anxious about losing it. Weiner's analysis illuminates this dynamic of reciprocity in TelaMaya, in which much is exchanged and much is held back. However, voluntourists and cooperative members did not participate equally in this dynamic. As the inheritors of Enlightenment conceptions of knowledge as a common good (Kuper, 1999), voluntourists tended not to view their knowledge as regional or culturally bound; instead, they saw what they knew as conforming to international professional standards and highly transferable. They were quite willing to share their skills and technical practices with others. Voluntourists' knowledge was universalising and not rooted to a particular place, and thus something they saw as a shared property to be freely circulated. By contrast, local indigenous knowledge has come to be seen, in part through development work, as a special intangible property held and cultivated by people. Joanna Davidson pointed out in the context of Guinea-Bissau that the current trend towards incorporating local knowledge in development projects, with the accompanying assumption that knowledge is 'extractable' and 'more knowledge is better' comes up against local attitudes towards knowledge as something to be kept and managed (2010, p. 2013). This article argues that cooperative members managed information, causing friction with voluntourists who viewed knowledge as something to be shared.

One of the primary functions of organisations like TelaMaya is to produce, disseminate, translate, and consume knowledge, to create connections between people and groups across international or micro-cultural boundaries (Naples & Desai, 2002). The voluntourists and cooperative representatives brought together in the TelaMaya volunteer programme occupied this role as brokers of meaning, with the voluntourists representing communities of consumers from their countries of origin and TelaMaya leaders representing their constituent weaving groups. The literature suggests that while NGO workers serve as brokers of meaning, they are not necessarily filling a pre-existing gap between the incommensurable discourses of development professionals and local people—instead, they are successful at convincing others of the importance of their information, creating spaces for themselves to accomplish their goals (Hilhorst, 2003; Richard, 2009; Sharma, 2006).

For some time, development organisations have sought to elicit and incorporate indigenous knowledge like that of TelaMaya members in their programmes, seeing this as a more participatory and 'bottom-up' approach than universalising, 'top-down' technocratic solutions. Scholars have critiqued the notion that Western and indigenous knowledge are dichotomous (Lewis & Mosse, 2006); however, it may be more useful to investigate how and when people invoke the idea of a binary between Western and indigenous knowledge (Yarrow, 2008). Examining how people claim knowledge or the ability to translate between fundamentally incompatible worldviews points to how the cooperative leaders (administrative officers and weaving group leaders) and volunteers needed to justify their value as intermediaries and their role in translating between divergent epistemologies. They used discourses of development strategically to position themselves as those with knowledge and resources and those they helped as lacking something. They defined their roles in relation to each other through the lens of their different bases of knowledge: one rooted in the local context, and one that travels.

Methods

This article is drawn from a wider study involving a total of 20 months of ethnographic fieldwork in the weaving cooperative office between July 2010 and January 2013, in two 3-month preliminary visits and one continuous 14-month visit. The primary mode of data collection was participant observation, based in the work taking place in the central office during the work week, including meetings with volunteers and clients, weaving classes and artisan fairs, and the filming of several documentaries, as well as excursions to visit the weaving groups with the voluntourists and independently. I also conducted semi-structured interviews in English and Spanish with 23 volunteer tourists and 5 cooperative leaders. In this article, I focus on the perspectives offered by two organisational representatives who were charged with the daily operation of the cooperative: the cooperative president, María, a woman in her late 50s from the Mam community of San Martín Sacatepéquez in Quetzaltenango, and the vice president, Roxana, a woman in her 30s from the Kaqchikel community of Pujujil II in Sololá.

Since 1988, TelaMaya has provided a source of income to hundreds of women in 17 weaving groups in 5 states in the western highlands of Guatemala: Sololá, Huehuetenango, Sacatepéquez, Quetzaltenango and Quiché. Founded with the financial and logistical assistance of Dutch foreign aid after the worst years of the Guatemalan Civil War (1960–1996), it has been independently member-run since 1995, with the support of approximately 50 international volunteers per year. Of the 99 volunteers I met during my fieldwork, approximately half were from North America (57) and a third from Europe (34), with smaller numbers of volunteers from Australia, Asia, or Latin America. The majority of volunteers were women under the age of 30 (65), and the second-largest category comprised men under the age of 30 (15). They were also overwhelmingly female (80). The cooperative helps 400 rural weavers access markets for their products by operating a retail shop for tourists in Quetzaltenango, the second-largest city in Guatemala, and exporting products to Fair Trade shops in the US, Sweden, Germany, and Australia. In addition to selling woven products, TelaMaya operated a weaving school to subsidise its operating costs, in which the cooperative leaders and volunteers taught foreign tourists the basic steps in backstrap weaving. Voluntourists provided translation services, staffed the shop, consulted on product design and marketing, reshaped the physical space of the office to adapt it to tourists' expectations, and managed the cooperative's online presence and communications with wholesale clients.

A pair of Swedish volunteers who worked with the cooperative in 2003 and returned to visit in 2012, Ana and Erik, said that they were impressed by the 'professionalisation' of the cooperative and its leaders, as evidenced by the increased standardisation of the store hours and products and capacity to fill large export orders. They credited the volunteers with having convinced the cooperative leaders to keep the store open during its advertised hours, maintain standard sizes and colours in their products, and present a limited number of examples of each product in the store rather than displaying the entire stock. The volunteers were probably a major factor in this evolution, as evidenced by moments such as when María cited the words of a specific former volunteer when she chided Roxana for allowing her children to leave toys in the office.

Tourists as sources of information

In describing her attitude towards the volunteer tourists her cooperative hosts, Roxana stated that each person has a gift, and that they are complementary: 'You [volunteers] speak Spanish and also English, because I don't speak English. I speak Spanish, Kaqchikel, and K'iche' because you don't speak them. God puts us together because we can help each other'. In this statement, Roxana framed the dynamic of voluntourism as one in which people from different geographic locations and knowledge communities come together to share their place-based knowledge; she presented it as a reciprocal, relatively egalitarian interaction, despite the many disparities between these groups. While TelaMaya leaders consistently utilised this rhetoric of reciprocity, this article will show that the reality of these information exchanges is far more complex.

Voluntourists were a valuable source of information about distant systems of value to the TelaMaya leaders, communicating their understandings of what clients look for in their products and representations of Mayan traditions, as well as their ideas about how small enterprises should operate. For instance, a Polish volunteer, Katarzyna, suggested that a fanny pack made of ethnic fabric would be wildly popular in Eastern Europe, and a South African volunteer with a fashion background proposed sewing two scarves together into a tunic. In an interview, Roxana stated:

> Volunteers are very important because they know their people, and since they're foreign volunteers, they know about foreigners' tastes ... what it is that they like: for example with the colours, the bag designs, how they like to see a place look.

Roxana noted that some of the products that did not sell well, including the placemats from her home weaving group, might not fit with foreigners' aesthetics: 'Maybe they don't like the colours because they're more ours. I know that in the United States there's autumn, winter, summer ... some products have gone out of fashion'. The officers and representatives took volunteers' aesthetic opinions very seriously, consulting them about patterns and colours that were popular in their home countries.

Volunteers shaped how TelaMaya presents its products aesthetically. When asked about the influence of volunteers in TelaMaya, Roxana gestured around the office to the murals and photos lining the walls, the displays in the shop and the weaving school. Voluntourists shared their insights as shoppers into the ways of arranging space and presenting products in the storefront in Quetzaltenango that would allow them to command higher prices. They worked to adapt it to the needs of visitors, making the public spaces more inviting and recognisable as a tourist destination. Volunteers frequently suggested that the store layout was too cluttered, commenting that the presentation of the products piled in baskets, like in the outdoor markets, led customers to expect low or flexible pricing. In December of 2012, a Dutch volunteer and the volunteer coordinators, a Canadian couple, painted the walls in lighter colours and rearranged the displays and products, placing many into storage to open up the space.

Voluntourists worked with Roxana to create a 'weaving museum' to enhance visitors' experiences and make the central office more of a tourist destination. Because TelaMaya was not Fair Trade certified, one of the goals of the museum was to make the visual argument that the organisation met certain Fair Trade standards, with a display showing the percentage of the sale price that went directly to the weaver. The voluntourists advised the officers on how to use contextual markers to designate the TelaMaya store as a 'fair trade' space, including photographs of some of the members, displays about the organisation's history and mission statement, and a standard store layout with price tags and labels to make tourists comfortable purchasing. These efforts at reframing the cooperative extended to its online presence as well: a volunteer who worked as a marketing consultant in the US, Jacqueline, commented that to command higher prices, TelaMaya would need to break out of the category of 'hippie/ethnic' backpacker souvenirs, and that one of the ways to do that was to photograph the products in Western-style environments with other luxurious objects.

Voluntourists also informed TelaMaya members about the kinds of stories and images they found appealing; in many cases, this included information about Mayan symbols or ways of life. As voluntourists sought to use quasi-anthropological information to add value to TelaMaya's products, the end result sometimes reflected their touristic expectations. Their mode of depicting Guatemalan handicrafts fit with Hendrickson's (1996) analysis of how international catalogues tended to represent 'Mayan' or 'Guatemalan' products: descriptions exoticised the products and emphasised the traditional processes used to hand-craft them. When TelaMaya's volunteers added new products to the online catalogue, they asked president María for information for the new captions: 'What do the designs in the weaving mean? What would this product have been used for traditionally?' At first, she was at a loss: 'This is a new item we started making. It doesn't mean anything; it's just something the weavers thought tourists would like'. Then inspiration struck María: 'You could just say that the designs are based in nature'. A volunteer took up this idea and added, 'And we could say that it is made using ancient backstrap loom techniques. That's not a lie'. In this and other instances, the process of (re)constructing the kind of Mayan heritage that tourists might find appealing represented a collaboration between the Mayan leaders and the voluntourists. Voluntourists in TelaMaya have become both consumers and producers of touristic imagery.

In addition to serving as consultants on product design and presentation, in both tangible and intangible terms, voluntourists sought to guide cooperative leaders with their knowledge of Western strategic planning and project management practices. However, the cooperative leaders were less receptive to these suggestions than other forms of information. In response to an interview question about what role volunteers should play in TelaMaya, Polish volunteer Katarzyna said:

> Add new points of view, like the things we learn from our countries – some other ways to, for example, promote [the products] – but I think that nobody is interested about our experience. I feel the problem, and I spoke to other girls, that the problem in Guatemala and also in the South American countries, is that they feel they know the best, and I think that that is the reason that they are developing so slow.

John, a young volunteer coordinator from the US, worked with the officers of TelaMaya and a wholesale US client to develop a long-term strategy for the organisation. He told the client:

> The ladies currently don't think about business and growth the same way we [Americans] do. It seems like their idea of growth is doing the same thing they're doing now on a larger scale … I think the lack of suggestions [for future development] comes from a lack of understanding about how things work on the other side of the business relationship.

John identified a need for the cooperative leaders to develop a more cosmopolitan acquaintance with ways of doing business in their clients' home countries. He claimed they had 'not yet grasped the idea' that long-term development could be funded through profits: 'They seem to have the mind-set that any additional projects must be covered by donations and outside money. I'm working to instil the idea that TelaMaya can increase profits and use that money to reinvest in the business'. John wanted to shift the officers' conception of their organisation from being 'aid'-focused to 'trade'-focused, a more appropriate neoliberal attitude. John emphasised the fundamental difference between his conception of doing business and that of the cooperative leaders, and articulated the need for voluntourists like him to reshape these ideas.

Similarly, a young social entrepreneur, Charmaine, wrote in her personal travel blog about her efforts to design a more organised-looking computerised spreadsheet for the cooperative to track inventory and sales:

> At times it is frustrating working in a developing country because things are so disorganized, especially in this organization. The women don't want the company to grow yet they want to make more money. They won't let people look over their finances (understandable) but still. Add to that an organization based on volunteers who have different skills, levels, and time commitments and you are starting to see what we are working with.

The officers told her they would not use a digital system without keeping a hard copy because they had lost files when a previous volunteer helped them transfer their system online; they asked her to help with the biannual store inventory instead. She later blogged that her time at TelaMaya had taught her 'to do what people need vs. what I think they need and want to do'. She claimed to be humbled because she had 'learned that what people need and are actually going to use will be much more useful than trying to do things my way. They operate differently than I am used to and I have learned to respect that'. Charmaine's realisation fits within an established genre of writing on development work, in which a young person's practical experience leads to grounded insights and a humbler attitude towards Euro-American interventions. Though this form of self-realisation was glib, the experience did seem to teach her about working in a different cultural context. The friction that voluntourists experienced working with TelaMaya occasionally challenged their approaches to development, their inclination to attempt top-down strategic planning giving way to more responsive approaches.

At the intersection between cultural tourism and development work, TelaMaya voluntourists were ambivalent about Guatemalan knowledge. In their applications to volunteer and semi-structured interviews, voluntourists described learning about Mayan culture as one of their main attractions to the cooperative. US applicant Katelyn's statement was typical:

> I hope to be able to contribute what I can to the cooperative while learning about the weaving traditions of Guatemala and the women who keep them alive. I am looking to volunteer because I hope to become immersed in the culture in a way that merely travelling through the area would not provide.

However, development work has historically had a different approach to local knowledge. Under the influence of the ideology of modernisation, many development workers initially saw their mission as one of combating non-Western practices, overcoming local barriers to the adoption of new technologies and standards in the service of homogenisation (Simon, 2006). This attitude persists in many important ways. While few voluntourists would explicitly claim that the local knowledge that drew them to visit Guatemala was an obstacle to development, they did tend to try to remake TelaMaya in the image of business models from their homes. When these efforts fell short, they became frustrated: 'In Guatemala people just want to do it and they want to keep doing it the same way but companies change and fashions change and so it has to be changed', observed Belgian volunteer Katrien.

In the face of the need for volunteers and their expertise, due more than anything to their dependence on the Fair Trade market and foreign clients, María made the claim for the importance of her own territorial expertise and knowledge. She addressed a volunteer meeting, saying:

> It's very important to speak English, it's the main thing [laughs bitterly] and it's what we know least about. But like I say to the *compañeras*, 'You shouldn't feel bad if you don't know how to speak Spanish, if you only speak your languages; don't feel bad because you are masters of weaving'.

María recognised that, in a world that has valorised literacy, technical skills, flexible thinking, and transferrable knowledge over more rooted, time-intensive practices, not all forms of knowledge are created equal, and asserted her identity as an artisan: 'It's part of our culture, the designs and the backstrap weaving'. Once, María explained to visiting weaving group representatives that Roxana was not in the office because she was at a training session in computing. María continued, 'On the other hand, I'm an artisan. I am very stupid and these things do not stick with me'. When Roxana sought to learn a skill dominated by volunteers, María reasserted the distinction between their local knowledge and the foreign knowledge that computing represents. Her association of computing skills with foreignness was a way of defining her role within the cooperative that may have been partly tactical, to avoid the expectation that she learn these skills, but was also part of the broader process in which these groups linked their skills and their identities.

Weavers as (occasionally reluctant) sources of information

TelaMaya members were ambivalent about the weaving school the organisation operated: it provided the cooperative with much-needed funds, but the weavers feared that they were reducing their market base by teaching *gringos* to weave. Roxana told me that a Mayan woman who did not belong to the cooperative had confronted her, asking, 'Why are you teaching the *gringos* to weave? You fools show the *gringos* how to weave and they're going to go back to the US and make their own businesses, and they're never going to buy from us again'. TelaMaya's leaders trusted that most weaving students would not retain enough of the process to replicate it at home. María hesitated when a German design student wanted to spend three months in an intensive weaving apprenticeship, because she felt that the student would learn too much about weaving techniques and patterns. María suggested that she and Roxana were relatively unconcerned about the threat of competition from a few tourists, as the cooperative members with the most contact and familiarity with the tourists. However, the rest of the board of directors thought differently and some of them were *celosas* (jealous/possessive) of their traditional knowledge:

> It's because they're worried that they're going to steal their culture. Look how they reacted when we started to teach students to weave. How they scolded us! They said, 'it's fine for you, because you're teaching in the weaving school, but it damages us because the students are going to set up their own weaving schools and not order from us anymore'.

This quote illustrates the internal divisions created by knowledge management, that other cooperative members were aware that María and Roxana had positioned themselves as bearers of weaving knowledge and were worried that they were taking advantage of this position for their personal gain.

A narrative I heard repeatedly during my time with TelaMaya was that volunteers and weaving students were studying backstrap weaving not to have a touristic cross-cultural experience but to steal Mayan women's traditional knowledge and set up their own weaving schools. María sighed, 'That's what makes us sad, because people take advantage of our *humildad* [poverty/humility], take advantage of the trust that we give them'. In particular, the board of directors accused a former volunteer coordinator from the US of exploiting the free weaving classes they gave her in exchange for her year of volunteer service to start a weaving school in Colorado. A former weaving student from the US, Cynthia, was said to have created a 'Mayan weaving school' in Brooklyn. I found Cynthia's website, which did feature photos of TelaMaya members and imply that weavers were receiving donations from the school, and showed it to TelaMaya's officers. They were angered that Cynthia claimed to have created partnerships and fostered relationships with Guatemalan weavers, when they had never heard from her again after she took a weaving class. They felt their images were being used to sell a false connection with TelaMaya. When I tracked Cynthia down to ask her about what had happened, she asserted that it was a misunderstanding, that she had indeed taken a weaving class at TelaMaya during her trip to Guatemala and taken photographs of the weavers, but that her partners were Mayan women living in New York.

The TelaMaya leaders told me that they had heard from volunteers that there were many backstrap weaving schools across the US. In the face of what they saw as a serious and growing threat of knowledge theft, Mayan women held to the idea that they were the only ones who could provide an authentic experience to tourists by virtue of their ethnic identities, which could not be appropriated. Weaving is one of the most important and visible markers and expressions of the cooperative members' ethnic, gender, and age identity (Hendrickson, 1995). Berlo (1991) argues that textile production, food production, and reproduction intertwine to define indigenous Guatemalan women's identity; Maya women use anatomical words for their looms, metaphorically linking weaving to giving birth. Women have traditionally clothed themselves with their weaving. Depending on the knowledge of the viewer, seemingly minute or subtle characteristics of the dress, ranging from the colour scheme or motifs within the cloth to the garment shape or wrapping technique, can signal the wearer's geographic origin, social status and role, and personality. Distinctive dress styles can indicate the wearer's gender, marital status, age cohort, ethnicity, linguistic community, socioeconomic status, membership in community organisations, and ceremonial roles (Pancake, 1991). The centrality of weaving to Guatemalan culture could (and does) occupy volumes but the important point here is that this knowledge is intimately connected with the cooperative hosts' ways of being. As María commented, 'I think that maybe some gringos do not want to pay for their flights and learn from other *gringos*, but I think that the majority prefer to learn from the actual indigenous women.' This claim suggests that, while they taught tourists to weave, they saw this knowledge as inalienable and non-transferrable because of its inextricable connection with their ethnic identity. Despite their confidence that voluntourists could only ever produce a faded imitation of the experience of learning to weave at TelaMaya, the cooperative leaders and the groups they represented worried that the foreigners in their target audience might not be sophisticated enough as consumers to recognise this.

Like the cooperative leaders, most tourists viewed their weaving classes as a touristic place-based experience connected to their visit to Guatemala, and not a skill that they would continue practicing in their home countries. 'When I travel I like to do things that I couldn't do at home, like learning to make curry from an Indian woman. I mean, I could make a curry at home, but it wouldn't be the same', declared one student. Another weaving student gave this typical list of reasons for taking classes at TelaMaya: 'I wanted to be productive and have a gift at the end. I wanted to do something different, not just buy from the *mercados*. It's so different from what I normally do'. She discussed the class as a way to pass the time in Quetzaltenango and create a souvenir, rather than a skill that she was planning to develop further.

When the volunteers heard that their cooperative hosts were concerned that *gringos* would learn how to weave for themselves and stop buying textiles, they laughed. One German textile design student, Anna, pointed out, 'It's available in YouTube videos anyway'. 'We're not giving away your secrets', agreed US volunteer Emma. In addition to rejecting the idea that knowledge of backstrap weaving was restricted to Mayan communities, volunteers from the global North rejected the idea that any tourists would go into production for their own use or for commercial gain. In Emma's words, 'I would say that approximately 0% of the people would go into business'. In part, they were rejecting the idea that selling weavings or weaving classes in the US would be as significant an industry as it would be in Guatemala, given their awareness of the opportunities open to women in the US labour market. On another level, the volunteers dismissed the weavers' concerns because they did not consider backstrap weaving a proprietary form of knowledge, as so frequently happens when communally held indigenous knowledge encounters a system designed to protect individual inventions.

In addition to their concerns about losing their livelihood to voluntourists, the cooperative hosts feared other kinds of intangible thefts. María announced offhand one day that volunteers frequently steal USB drives from TelaMaya in order to get information about TelaMaya as an institution: 'They say that the USB got stolen but it's a lie, they've taken it. [...] It bothers me that the volunteers always take away TelaMaya's information, and how do they use it?' María claimed that they caught a former volunteer coordinator in the office after work hours going through their papers. The incident led them to distrust him, for seeming to covertly gather information about their business. According to María, 'He told me, "If you trust me, what's the problem? Is it because I'm a foreigner that you don't trust me?"' I pressed María about what kinds of 'information' he might have been seeking. She felt that he was trying to find data to discredit the cooperative, because he did not believe that the cooperative had as many members as it claimed.

The officers claimed to be happy to share information about the cooperative with volunteers, because they knew what volunteers could do to help them. However, in practice they were reluctant to give out important institutional facts. A volunteer wanted to interview María about the cooperative's exports to understand its market position. She balked at the idea: 'But why does she want to know that? That's something that's very much ours. The women [of the cooperative] do not want to say everything about the cooperative's business to just any person'. Roxana added, 'The women are very strict with us. They say, "Yes, let's share everything, but how will they help us? Why do they want this information?" Sometimes people come and get our information and don't do anything for us'. When a professor of social work from the US interviewed Roxana and María about TelaMaya's business practices and fiscal situation, they were evasive, refusing to answer some questions and claiming not to know the answers to questions about income and expenses, finally admitting, 'We can't say exact quantities because that's private'. He asked, 'What types of problems are the weavers facing?' María's answer was curt: 'None. They're already trained in weaving'. These kinds of evasions and half-truths protected the information and knowledge that the cooperative leaders considered '*muy nuestra* [very much ours]'.

Conclusion

One of the most interesting aspects of the TelaMaya volunteer programme is that it put voluntourists in the position of working in the tourism industry, attending to customers in the store and marketing the cooperative to other tourists as a destination for shopping, weaving lessons, and volunteering. Tourists, as volunteers, became intimately engaged with the process of representation in a way that disrupted classic dichotomies between 'guest' and 'host'. One of the factors accounting for the differences between the attitudes of voluntourists and their hosts towards sharing their knowledge may be their different structural position within volunteer tourism. While the presence of tourists is constant, individual representatives come and go in an endlessly renewed flow; by contrast, the host population (the cooperative leadership) is relatively fixed. Each tourist spends a relatively short time doing voluntourism, whereas for their hosts, it is a way of life. It may be easier to maintain an attitude of openness and generosity over the short term. Cooperative members have grown weary of providing the same explanations and lessons to each new cohort of volunteers, as the returns on the information they have offered are not always immediately forthcoming.

In their daily interactions with voluntourists, TelaMaya leaders revealed their ambivalence about the knowledge exchanges entailed in ethical consumerism (voluntourism and Fair Trade). They wanted to benefit from their stories and images by participating in international Fair Trade markets. However, they also wanted to protect their proprietary knowledge both as an inalienable form of cultural heritage and a market resource. The members of TelaMaya have staked a claim for worth on a global scale based in their traditional Mayan knowledge, the value of which is evidenced by the desire of foreign tourists and buyers across the globe to consume it. Having based their importance to the organisation on their proprietary knowledge, the cooperative members and particularly those involved in the cooperative administration felt the need to defend it and their right to be the ones distributing this knowledge. Mobility of outlook and practice is not necessarily simply an advantage; it can also be a risk, as forms of knowledge once rooted in a place become mobile and potentially ripe for theft. They were concerned that their knowledge might be a limited resource, something that, once shared, is gone past recovery.

The 'critical turn' in tourism studies, which focuses on shifting imaginaries and emancipatory ideals, has been criticised for failing to be 'critical' of power disparities because it directs attention away from the economic and political effects of tourism (Bianchi, 2009). By bringing theories about the development of cosmopolitanisms in tourism into conversation with theories about cultural knowledge as a limited resource, this article helps to address this myopia towards material inequalities. It draws together the critical turn in tourism studies with a grounded analysis of the daily frictions and material issues of the hosts of tourists. It unites the intangible shifts in consciousness represented by cosmopolitanism with the notion that, in many ways, the global traffic in knowledge has made intangible goods, the most crucial to people's political and economic realities.

Acknowledgements

I wish to thank Françoise Dussart and Samuel Martínez for their guidance and support, as well as my colleagues from the Guatemala Scholars Network; the Inter-American Foundation; the UConn Human Rights Institute; El Instituto; the UConn Department of Women's, Gender, & Sexuality Studies; and the UConn Department of Anthropology. The constructive comments made by three anonymous reviewers also contributed to improving this paper.

Disclosure statement

No potential conflict of interest was reported by the author.

Funding

This work was supported by the Inter-American Foundation and the University of Connecticut.

ORCID

Rebecca L. Nelson http://orcid.org/0000-0003-0347-0043

References

Appiah, K. A. (2006). *Cosmopolitanism: Ethics in a world of strangers*. New York, NY: W. W. Norton.

Berlo, J. C. (1991). Beyond bricolage: Women and aesthetic strategies in Latin American textiles. In M. Schevill, J. C. Berlo, & E. B. Dwyer (Eds.), *Textile traditions of Mesoamerica and the Andes: An anthology* (pp. 437–479). New York, NY: Garland.

Bianchi, R. V. (2009). The 'critical turn' in tourism studies: A radical critique. *Tourism Geographies: An International Journal of Tourism Space, Place and Environment, 11*(4), 484–504.

Brown, M. (2003). *Who owns native culture?* Cambridge: Harvard University Press.

Bryant, R. L., & Goodman, M. K. (2004). Consuming narratives: The political ecology of 'alternative' consumption. *Transactions of the Institute of British Geographers New Series, 29*(3), 344–366.

Butcher, J. (2003). *The moralisation of tourism: Sun, sand ... and saving the world?* New York, NY: Routledge.

Butler, R. (2004). Alternative tourism: The thin end of the wedge. In S. Williams (Ed.), *Tourism: Critical concepts in the social sciences* (vol. 4, pp. 310–318). London: Routledge.

Clifford, J. (1992). Traveling cultures. In L. Grossberg, C. Nelson, & P. Treichler (Eds.), *Cultural studies* (pp. 96–116). London: Routledge.

Davidson, J. (2010). Cultivating knowledge: Development, dissemblance, and discursive contradictions among the Diola of Guinea-Bissau. *American Ethnologist, 37*(2), 212–226.

Ferguson, J. (1999). *Expectations of modernity: Myths and meanings of urban life on the Zambian copperbelt*. Berkeley: University of California Press.

Gibson-Graham, J. K. (2006). *A postcapitalist politics*. Minneapolis: University of Minnesota Press.

Hannerz, U. (2006). *Two faces of cosmopolitanism: Culture and politics*. Serie: Dinámicas interculturales Número 7. Barcelona: CIDOB.

Hendrickson, C. (1995). *Weaving identities: Construction of dress and self in a highland Guatemala town*. Austin: University of Texas Press.

Hendrickson, C. (1996). Selling Guatemala: Maya export products in U.S. mail-order catalogues. In D. Howes (Ed.), *Cross-cultural consumption: Global markets, local realities* (pp. 106–121). New York, NY: Routledge.

Hilhorst, T. (2003). *The real world of NGOs: Discourses, diversity, and development*. London: Zed Books.

Hudson, I., & Hudson, M. (2003). Removing the veil? Commodity fetishism, fair trade, and the environment. *Organization & Environment, 16*, 413–430.

Kuper, A. (1999). *Culture: The anthropologists' account*. Cambridge: Harvard University Press.

Lewis, D., & Mosse, D. (2006). *Development brokers and translators: The ethnography of aid and agencies*. Bloomfield, CT: Kumarian Press.

Lyon, S. (2006). Evaluating fair trade consumption: Politics, defetishization and producer participation. *International Journal of Consumer Studies, 30*(5), 452–464.

Macleod, D. (2004). Alternative tourism: A comparative analysis of meaning and impact. In S. Williams (Ed.), *Tourism: Critical concepts in the social sciences* (Vol. 4, pp. 189–205). London: Routledge.

Naples, N., & Desai, M. (Eds.). (2002). *Women's activism and globalization: Linking local struggles and transnational politics*. New York, NY: Routledge.

Notar, B. E. (2008). Producing cosmopolitanism at the borderlands: Lonely planeteers and 'local' cosmopolitans in Southwest China. *Anthropological Quarterly, 81*(3), 615–665.

Pancake, C. (1991). Communicative imagery in Guatemalan Indian dress. In M. Schevill, J. C. Berlo, & E. Bridgman Dwyer (Eds.), *Textile traditions of Mesoamerica and the Andes: An anthology* (pp. 45–62). New York, NY: Garland.

Princen, T., Maniates, M., & Conca, K. (Eds.). (2002). *Confronting consumption*. Cambridge: MIT Press.

Richard, A. (2009). Mediating dilemmas: Local NGOs and rural development in neoliberal Mexico. *Political and Legal Anthropology Review, 32*(2), 166–194.

Salazar, N. B. (2010). Tourism and cosmopolitanism: A view from below. *International Journal of Tourism Anthropology, 1*(1), 55–69.

Sharma, A. (2006). Crossbreeding institutions, breeding struggle: Women's empowerment, neoliberal governmentality, and state (re)formation in India. *Cultural Anthropology, 21*(1), 60–95.

Simon, D. (2006). Separated by common ground? Bringing (post)development and (post)colonialism together. *The Geographical Journal, 172*(1), 10–21.

Szerszynski, B., & Urry J. (2006). Visuality, mobility and the cosmopolitan: Inhabiting the world from afar. *The British Journal of Sociology, 57*, 113–131.

Taylor, P. L. (2005). In the market but not of it: Fair trade coffee and forest stewardship council certification as market-based social change. *World Development, 33*, 129–147.

Vertovec, S., & Cohen, R. (Eds.). (2002). *Conceiving cosmopolitanism: Theory, context and practice*. Oxford: Oxford University Press.

Vrasti, W. (2013). *Volunteer tourism in the global south: Giving back in neoliberal times*. New York, NY: Routledge.

Wardle, H. (2000). *An ethnography of cosmopolitanism in Kingston, Jamaica*. New York, NY: The Edwin Mellen Press.

Weiner, A. (1992). *Inalienable possessions: The paradox of keeping-while-giving*. Berkeley: University of California Press.

Werbner, P. (1999). Global pathways: Working class cosmopolitans and the creation of transnational ethnic worlds. *Social Anthropology, 7*, 17–35.

Yarrow, T. (2008). Negotiating difference: Discourses of indigenous knowledge and development in Ghana. *Political and Legal Anthropology Review, 31*(2), 224–242.

Yúdice, G. (2003). *The expediency of culture: Uses of culture in the global era*. Durham, NC: Duke University.

There's a troll on the information bridge! An exploratory study of deviant online behaviour impacts on tourism cosmopolitanism

Aaron Tham and Mingzhong Wang

ABSTRACT
Existing literature has portrayed tourism as a conduit for learning and transformation, with a view that such mobilities elucidate tourism cosmopolitanism. While such propositions are insofar useful in characterising global dispositions, very little has been done to investigate how the virtual space has helped to develop or curtail effects of tourism cosmopolitanism. The virtual space has transformed tourism engagement and consumption patterns, though few studies have paid attention to deviant online behaviour and its corresponding effects on cosmopolitanism. Drawing from outcomes of trolling instances within forums, this paper argues that deviant online behaviour is of consequence as an antecedent in determining one's disposition to destinations, challenging the decision type and raising serious concerns to self-image and virtual representations. These attacks on any individual necessitate a cautious approach for scholars and practitioners when engaging with the electronic realm as a different, yet instrumental facet in characterising tourism cosmopolitanism.

Introduction

The notion of tourism as a vehicle for learning and transformation has been well documented in literature (*inter alia* Cuffy, Tribe, & Airey, 2012; Falk, Ballantyne, Packer, & Benckendorff, 2012; Mitchell, 1998; Schmelzkopf, 2002). Aside from being of hedonic motives, scholars espouse the value of tourism in developing a nuanced perspective of the world, in the hope of bringing cosmopolitan outcomes to fruition (Pritchard, Morgan, & Ateljevic, 2011; Swain, 2009). The aspirational goals of attaining learning and transformational outcomes through tourism have likewise been identified through activity types, such as volunteer tourism (Coghlan & Gooch, 2011), field trips (Xie, 2004) and backpacker travel (Pearce & Foster, 2007). While these types of travel arrangements are likely to possess some learning orientations, other scholars have questioned whether tourism experiences are instrumental in shaping cosmopolitan thinking and behaviour (Hollinshead, Ateljevic, & Ali, 2009; Johnson, 2010). These concerns are made on the back of tourism characterised as being short term (Sin, 2009), and of a discretional nature (Thompson & Tambyah, 1999).

These criticisms notwithstanding, tourism has witnessed significant growth in terms of tourist mobilities (Hall, 2011). The movement of tourists across places and time provides a rich landscape upon which cosmopolitan principles are sown (Bianchi & Stephenson, 2013). In a broad sense, tourism can be encountered both in the 'real' world, and in the 'reel' (Alderman, Benjamin, & Schneider, 2012; Lee, 2012). By the same token, cosmopolitan thinking and behaving can likewise be concocted when engaging with the physical, and the virtual (Hannam, Butler, & Parris, 2014). While a body of literature surrounds tourism cosmopolitan in the physical realm, very little exists to characterise the virtual cosmopolitan space in tourism (Tavakoli & Mura, 2015). The extant literature concerning tourism digital cosmopolitanism has argued that the online realm characterises what Narayan (2013) terms a networked culture. Such cultures lend a different lens to the dominant discourses surrounding cosmopolitanism because they operate in across borderless territories (Goode, 2010). As such, networked cultures reframe cosmopolitanism principles by facilitating knowledge exchange without the necessary physical mobilities. In addition, Mills and Green (2013) postulated that virtual mobilities can steer one's dispositions towards beliefs and preferences. Applied to tourism, digital cosmopolitanism can be agents of change towards destinations, cultures and travel motivations. For instance, Rokka and Moisander (2009) found that online communities ascribed greater ecological awareness and behaviour among global travellers.

However, online fake reviews and deviant behaviours also cast negative impacts on the tourism industry (Schuckert, Liu, & Law, 2016). McCosker (2014) raised the issue of trolling behaviour in online communities, and highlighted that technological innovations may not always lead to social progress or positive connections between people with different perceptions and beliefs. Although some studies (see Filieri & McLeay, 2014; Hernandez-Mendez, Munoz-Leiva, & Sanchez-Fernandez, 2015; Sparks, Perkins, & Buckley, 2013) have tried to examine how online reviews may affect traveller's decision-making and destinations' reputation, there is limited investigation on the phenomenon of online trolling behaviour by tourism actors. The aim of this paper is to elucidate further knowledge about online behaviour, particularly deviant outcomes from trolling encounters, and its impact on tourism cosmopolitanism.

Background

Cosmopolitanism and the digital landscape of tourism

Although cosmopolitanism does not have a universally adopted definition (Beck & Sznaider, 2006; Lu, 2000), scholars are in overall agreement that cosmopolitanism can be coached to be more developed (Delanty, 2006). In other words, the individual undertakes meaning-making from an immersion with the less, or unfamiliar people, objects and places to grasp a global orientation (Ossewaarde, 2007). Often used interchangeably, global citizenship is used to describe individuals who are socially responsible, possess global skillsets and equipped with civic-mindedness (Morais & Ogden, 2011; Stoner et al., 2014).

The prevalence of the Internet triggered the study of digital cosmopolitanism, which focuses on the connection and collaboration among people in cyberspace to address linguistic, national or cultural barriers (Sobre-Denton, 2016). Table 1 illustrates the key differences between cosmopolitanism and digital cosmopolitanism.

Tourism has witnessed the explosive growth of digital trends in the last decade (Leung, Law, van Hoof, & Buhalis, 2013). The online domain has transformed how tourism-information is produced and consumed, where current trends show the rise in user-generated contents to a large audience almost instantaneously. Such developments have had positive impacts on tourists and practitioners. From the tourist perspective, tourism-related information can be easily and readily accessed to assist with information search and decision-making (Xiang & Gretzel, 2010). For tourism practitioners, social media offers a valuable tool to disseminate new products and services to a large potential market in conjunction with traditional collaterals (Zeng & Gerritsen, 2014). Collectively, the benefits of social media ensure that desired tourism experiences are produced, and 'pre-tested' in a timely and responsive manner.

However, the growth of social media has raised some concerns. For instance, some scholars have expressed doubts regarding the credibility of social media due to the ease of publishing contents online (Kusumasondjaja, Shanka, & Marchegiani, 2012; Mack, Blose, & Pan, 2008). As such, it is envisaged that any social media user will be expected to develop some heuristics to guide their engagement on social media and to assess sources or contents for credibility (Ong, 2012; Papathanassis & Knolle, 2011). As credibility is an antecedent to influence, a social media user is anticipated to involve cognitive appraisal of social media contents to make more informed decisions, as in the case of a highly experiential tourism environment (Scott & Orlikowski, 2012). Overall, the current scope of literature has delineated user profiles, motivations for social media use and the proposed influence on preferences and choice outcomes. Despite these developments, few studies in tourism have examined the effect of deviant social media behaviour in framing digital cosmopolitanism. The aim of this paper is to provide further insights to this knowledge gap.

Deviant behaviour

A broad definition of deviant behaviour is the display of actions or words that run contrary to the norms within any given setting (Dubin, 1959; Gibbs, 1966). Examples in a tourism setting include binge drinking (Smeaton, Josiam, & Dietrich, 1998), sex parties (Ryan & Kinder, 1996), vandalism (Haralambopoulos & Pizam, 1996) and

Table 1. Conceptual differences between tourism cosmopolitanism in a physical and digital environment.

Dimensions	Physical cosmopolitanism	Digital cosmopolitanism (through the lens of trolling)
Mobility	Through space and time	Based on one facet of virtual encounters
Consumption	Unhurried and self-paced	Expedited and immediate
Curiosity	Intellectual to becoming a more learned tourist	Superficial and fragmented
Risking encounters	With the goal of becoming resilient	Leading to ignorance or rebuttals
Mapping reflection	To reconstruct notions of place and people	That results in defending one's beliefs and attitudes
Openness	In accepting cultural diversity	Limited to others who are similar-to-me
Interpretation	Critiquing difference and building consensus	In selective liminal spaces

Source: Adopted from Swain (2009).

drug use (Uriely & Belhassen, 2005). These exemplars are considered deviant behaviours because they often leave a detrimental effect on other stakeholders and can cast some market segments or certain destinations in a very negative light. A common characteristic of these deviant behaviours is that such acts occur while the tourist is on-site at a destination. For this reason, some studies have focused on examining why deviant behaviours occur. Uriely, Ram, and Malach-Pines (2011) found that an unconscious drive exists among tourists wanting to misbehave outside the scope of their everyday lives. In contrast, de Albuquerque and McElroy (1999) contended that deviant behaviours exist because of peer pressure to undertake acts of mischief.

Clearly, deviant behaviours can exist for any reason and may be directed at people or places. While there is some literature to highlight deviant behaviour at a destination, very little is known about deviant behaviour that takes place prior to the destination visit. This paper aims to shed light on what deviant behaviours are manifested within social media communities, given that such spaces are often the initial points of interaction between a tourism decision-maker and the scope of information available. Specifically, the paper is concerned with how trolling on social media has impacted the tourism online community, and its corresponding effect on digital cosmopolitanism developments. The next section will review literature pertaining to the notion of trolling.

Trolling

Perhaps one of most well-known instances of a troll is in the children's fiction book: *The three billy goats gruff* (Brown, 1957). In the story, the troll is a devious monster who lives under a bridge and intercepts each of the goats attempting to cross the bridge in search of green pastures. The first goat tells the troll to let him through because the subsequent goat has more meat and would make a better meal. The second goat gives the same comment to the troll and eventually the third goat, because of his stature, overpowers the troll who eventually leaves the vicinity forever. While this story is a fictional tale, it nonetheless demonstrates the evil nature of the troll that potential victims attempt to avoid.

In this research, the term trolling refers to behaviour that is aimed at antagonising others through abusive words or graphics that detracts from the given norms of the social media community (Herring, Job-Sluder, Scheckler, & Barab, 2002; Hopkinson, 2013). Trolling often results in the 'troll' being evicted from the online community, though the direr consequences are the image of the social media site and the physical and mental distress caused to victims. As the characteristics suggest, trolling can be considered a form of cyberbullying (Buckels, Trapnell, & Paulhus, 2014; Mehari, Farrell, & Le, 2014). As the number of social media sites have increased in the last decade, trolling behaviours have likewise spiked (Bishop, 2013; Hardaker, 2010). Bishop (2014) postulated a typology of trollers comprising four types – *Haters, Lolcows, Bzzters* and *Eyeballs*. The distinctive characteristics of these four types are presented in Table 2.

In a tourism environment, the animosity caused by vitriolic comments can cause any social media user to exclude himself or herself from the community, and worse still decide not to undertake tourism at all. To complicate matters, trolls that have been evicted can easily assume another pseudonym or fabricate details to re-enter the community (Jane, 2014). The ease of entry to social media communities is a challenging one for policing troll behaviour because various countries have their own perspective on Internet censorship and regulations, though it is also highly probable that a troll could be anywhere on the planet (Yar, 2012). As such, trolling remains a highly problematic issue to mitigate and manage, as in the case of tourism. To the authors' best knowledge, very little is known about the prevalence of trolls within the tourism online communities, except for Mkono (2015). In her study, Mkono (2015) reports on instances related to trolling behaviour in tourism, and how forums like TripAdvisor respond to such deviant incidents. However, Mkono (2015) also highlighted the need to examine trolling and its impacts on individuals and their digital footprints. The paucity of literature characterising trolling in terms of tourism digital cosmopolitanism lends the justification to undertake this research. Derived from the literature on cosmopolitanism and trolling as a form of deviant behaviour has led to the development of the conceptual framework (Figure 1) for this study.

The research seeks answers to the main research question:

How do trolling incidents affect tourism digital cosmopolitanism developments?

Table 2. Trolling typology.

Haters	Individuals who inflame situations for no real benefit
Lolcows	Individuals who provoke others to turn attention to themselves
Bzzters	Individuals who enjoy chatting regardless of the accuracy or usefulness of comments
Eyeballs	Individuals who observe others for a chance at posting a provocative message

Source: Bishop (2014).

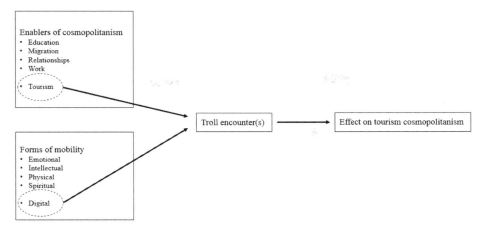

Figure 1. Conceptual framework.

The following secondary questions will help to address the main research question:

How does the interaction between trolling and the following characteristics impact on tourism digital cosmopolitanism in terms of:

- Individual characteristics such as age, gender, ethnicity, religion?
- Tourism decisions – for example destination, hotel, mode of transport, places of interest?
- Travel contexts (e.g. with a family, timing of travel, special interests)?

Answers to the research questions will help conceptualise digital tourism cosmopolitanism in an era of highly unmoderated social media contents. The managerial implications of the research are to steer social media users and other stakeholders to mitigate the negative impacts of trolling to re-position the tourism experience in a favourable manner.

Method

To obtain a less-intrusive nature of trolling behaviour on social media, a netnographic approach was undertaken by the authors. Netnography, proposed by Kozinets (2002), is a type of research method that investigates online contents which has been increasingly relevant to social media in a tourism context. For instance, Mkono (2011) utilised netnography to examine tourist experiences consuming specific cuisines in Africa. In contrast, Wu and Pearce (2013) applied netnography to better understand Chinese self-drive tourists during their vacations around Australia. According to Kozinets (2002), researchers can adopt a proactive stance by engaging within the online communities as a participant, or where the need arises, a passive stance as a 'lurker' who observes online narratives without intrusion on the subject matter. As this research may raise some elements of discomfort among the social media community, it was decided that having a lurker perspective was more suited to the intrusive nature of the investigation. This approach is consistent with the work of Bjork and Kauppinen-Raisanen (2012) who adopted netnography in unpacking the online discussions of risks among tourists regarding three destinations.

Guided by the principles of netnography, the researchers first identified appropriate forums that were likely to generate insights as to trolling behaviour. Two leading tourism forums were chosen for the research because of their widespread audience and depth of discussions regarding tourism matters around the world, which has been adopted by other scholars (Arsal, Woosnam, Baldwin, & Backman, 2010; Govers, Go, & Kumar, 2007; Vasquez, 2011). These forums were:

(1) TripAdvisor (http://www.tripadvisor.com.au/ForumHome) and
(2) Lonely Planet Thorn Tree (https://www.lonelyplanet.com/thorntree).

However, troll encounters that had surfaced within TripAdvisor threads appeared to be shut down by the site administrators or moderators once these have been brought to their attention. As no further details to these trolling incidents were made publicly available, the Lonely Planet Thorn Tree (TT) forum was utilised for this research.

In brief, the TT forum is located on the Lonely Planet website as a repository of online information disseminated by travellers for travellers (Dwivedi, 2009). The forum is divided into two main categories – country forums and interest forums. Prior to its inception, Lonely Planet is a company known for its travel guidebooks (Hwang, Jani, & Jeong, 2013). Its main target market were backpackers and low-cost travel, and hence the guidebooks were written with these audiences in mind (Dippelreiter et al., 2008). The history of the brand is likewise evident in the TT forum. Some scholars have noted the similar presence of backpackers on the forum (e.g. Iaquinto, 2012; Paris & Teye, 2010). This characteristic distinguishes the forum from other ones such as TripAdvisor, which caters to a broader tourist segment.

TT forum's association with backpacking is central to the research. This is because different scholars have alluded to the notion that backpacking is a form of tourism that has cosmopolitan trajectories (Bui, Wilkins, & Lee, 2013; Cohen, 2011; Molz, 2008). The physical movement of individuals across time and space, coupled with the stripping away of home comforts, requires backpackers to think and learn as to how to economise and make decisions regarding their travel itineraries (Ooi & Laing, 2010). It is also likely that backpacking journeys will bring travellers to host communities offering niche experiences as a proxy for cultural exchange (Walsh & Tucker, 2009). Hence, if it is accepted that backpacking and other low-cost travel are forms of tourism cosmopolitanism, then how does adding an additional digital lens through the Internet and social media moderate (un)desired outcomes? The paucity of information pertaining to this topic is the justification for this research to be undertaken.

Related to cosmopolitanism, the process in which a poster decides to place a question or comment in the forum suggests that he or she is receptive to others' commenting on their post, and possibly defending a viewpoint. This is what Aroldi and Colombo (2013) contend to be an entrée towards cultural appropriation, which forms the basis for digital cosmopolitanism to flourish. As the forum is an open platform where others can respond to any given post, it serves as a rich ecosystem for exchange of ideas and thereby fuel the appetites for digital tourism cosmopolitanism (Lewis, 2015).

To obtain relevant data specific to the research, a webcrawler was developed to capture threads which had the word 'troll' appear on the TT forum. The webcrawler was implemented with Python programming language. The webcrawler sent the query keyword 'troll' to the TT forum, and discovered 1996 threads. The search engine of the TT forum used fuzzy keywords to retrieve related threads. For example, for the query word 'troll', threads containing similar words, such as 'stroll' or 'toll' were also returned. Therefore, each thread was then downloaded and analysed to check the appearance of different variations of the word 'troll', such as 'trolls', 'Troll' and 'TROLL'.

The webcrawler identified a total of 382 occurrences (from October 2012 to August 2016) of the word 'troll' that formed the initial data set for analysis. Guided by the principles of identifying trolling types and behaviour as espoused by other scholars (Bishop, 2014; Hardaker, 2010; Hopkinson, 2013), the first 30 trolling incidents were coded into separately by two scholars to ascertain if the data were coded consistently. This process revealed that both scholars coded 22 trolling incidents into the same category, where the remaining 8 incidents were discussed and agreed upon as to their specific category. The two scholars then proceeded to code the remaining 352 incidents into categories. Each trolling incident was assigned to their respective poster to further assess how long each individual has been on the TT forum, and the number of posts made. On average, most posters were subscribers to the TT forum for over 7 years, and posted 4400 messages. These numbers lend further credibility that the posters were not novices to the forum, but instead are active members who have engaged regularly on various topics. The proactive nature of participatory web cultures is the ethos of digital cosmopolitanism practices, as supported by other studies (Boczkowski & Siles, 2014; Vasudevan, 2014). The authors have also been on the Thorn Tree forum as well as TripAdvisor for at least six years, regularly engaging in different threads to comment and solicit tourism-related information. As such, the authors were familiar with the TT forum conventions, which are located at: https://www.lonelyplanet.com/thorntree/community-guidelines.

The next section will present the findings and discussions in relation to the research questions. All the quotes presented in this research have been assigned a pseudonym to de-identify participants on the TT forum.

Findings and discussions

This section is divided into the follow sub-sections to help elucidate how trolling incidents can influence tourism digital cosmopolitanism. As an exploratory study, this research is not as much concerned about proving hypotheses but rather uncovering little known information about trolling and its effect on digital tourism cosmopolitanism. This assertion is built on the work of Mkono (2015) in extending the body of literature surrounding trolling in tourism, and illuminates how trolling alters the lens of digital tourism cosmopolitanism. This research also adopts a broad view of trolling as a subset of online deviant behaviour, which is consistent with how extant literature has portrayed trolling (see Bishop, 2014).

First, the section explains the categorisation results for the data set. Then, supported by some quotes, the section will illustrate how some social media users seek to glean more insights about a destination by presenting their proposed travel itinerary in the forum. This process is consistent with the work of Jacobsen and Munar (2012), who argued for social media to be an easy and convenient means to solicit tourism-related information. Next, the way troll identification is discussed, with the aim of exploring how online comments are linked to the notion of tourism cosmopolitanism thinking and behaving. Finally, the effect of trolling encounters on victims will be analysed considering tourism cosmopolitanism.

Troll categorisation

Each thread in the data set containing the word 'troll' was manually coded into one of the seven categories. The number of threads in each category is shown in Table 3.

About 44% of the threads tagged by forum users as 'troll' were in fact not related to deviant behaviour. Among trolling threads, 38% tried to target individuals or group of users, 24% were related to the image or reputation of the destination and 10% were about the travel features, such as transportation and backpacking. Another 37% of posts tried to use unrelated comments to hijack or disturb the communication.

Thread initiation

As with any tourism forums, a user may glean through existing threads to seek potential contents related to a destination or topic (Easton & Wise, 2015). However, users may also initiate new threads for discussion. Drawing from the data set where trolling behaviour was encountered are a few exemplars:

> I travel happily believing that I am way too old for anyone to be bothered to harass me, one of the perks of old age, I thought:) I have just returned from 2 weeks in Morocco... but when one young man touched me on the breasts and then the bum I really saw red, particularly as he seemed to be convinced that I would love to pay him for his sexual favours!! ... Have others had similar experiences? Which are the countries to be wary in? (Sharon)

> Myself and my girlfriend are contemplating a fortnights trip to somewhere in Greece. We're particularly interested in going somewhere that can give us a very subjective, but in-your-face attitude towards the present migrant/refugee problem... We'd like to talk to some of these folks about these situations. Any ideas where we should head to? (Sean)

> Hi, my husband and I are traveling to Costa Rica (Pacific Side) for a month this November to surf and soak up the sun. I have armpit hair and am wondering about any sort of cultural taboos I might be breaking if I don't shave. I can't seem to find anything online that says one way or another. Thanks! (Jessie)

As the three above-mentioned quotes show, thread initiation could be about any topic and appearing in different regions within the TT forum. These quotes lend some instances of presenting the poster's stance about a topic, and consulting other forum users about it. The different topics could be areas that one could read or learn from a range of channels, including forums on social media. However, for some of TT users, the forum appears to be a proxy to develop cosmopolitan thinking and behaving in the form of digital mobility. Yet, from other poster's perspective, there remains some caution as to what one can articulate, who may respond and how the response will appear:

> Hello! Me and two or three friends are planning on going to South America, rent ourself offroad bikes, and travel around the continent for about six weeks... What we are concerned about is the safety. I've tried googling a lot, but haven't got much wiser. Therefore I turn to the travelling experts here at lonely planet. Is there any specific countries or areas that we should avoid, or at least be extra careful in? (Matt)

> First of all, thank you for replying and taking my questions seriously. I know all my posts sound quit stupid, but this is my first time abroad without family... Next to rabies and landmines my last worry is that when flying home, some drugs will be planted in my luggage against my will. I will be flying home alone: Phnom Penh – Beijing – Schiphol – Brussels. I bought a backpack on purpose that I can fully close with zippers and will buy some locks too for the bags. But apparently those can be opened by the luggage handlers so there is not really a way to protect yourself from these things I guess... When you think rationally, it sounds really stupid of course. Can anyone confirm this is just an urban legend Or does anybody has tips / tricks to prevent this from happening? (Melissa)

The views portrayed by Matt and Melissa in starting new threads demonstrated signs of vulnerability, with

Table 3. Trolling incidents by category.

Category	Frequency
Person-centric	82
Destination-related	52
Trip characteristic	22
Political or religious issue	11
Advertisement	12
Nonsensical comment	37
Not considered trolling behaviour	166
Total	382

the comments appearing to seek the need for others to validate or to support their decisions within the respective forums. As tourism is a highly experiential encounter with little ability to pre-test the potential visit to a destination, the use of social media are a means to also reduce post-decision dissonance (Tanford & Montgomery, 2015). This vein of interaction is somewhat peculiar as other TT users may not possess source or content credibility to influence the thread initiator, though some of the users continue to engage with others despite their unknown identifies.

The digital form of tourism cosmopolitanism offered by social media is unique in this regard as it provides for a widely inclusive interactive platform (with any user able to participate with a username and password), though it presents some dimensional differences to other concepts surrounding cosmopolitanism. Literature surrounding cosmopolitanism espouses the principles of being immersed in lesser known cultures, beliefs and societies to gain a more nuanced understanding of how to live with, and respect others (*inter alia* Calcutt, Woodward, & Skrbis, 2009; Delanty, 2006; Roudometof, 2005; Sluga & Horne, 2010; Young, Diep, & Drabble, 2006). Yet, social media appears to embody cosmopolitanism that is aligned with the work of Ossewaarde (2007) who envisions cosmopolitanism amidst the presence of individuals not acquainted with one another. Moreover, the threads shown previously in this section are more akin to taking a stance and expecting others to accept their view, building on homophily ('similar-to-me') principles. This tone of establishing a person-centred mentality epitomises the critique of cosmopolitanism as a lofty aspiration, when in reality, some individuals are not as interested in dealing with differences of opinions (Beck, 2002; Lu, 2000; Roudometof, 2005). However, rather than considering cosmopolitanism as a dichotomous outcome of either being cosmopolitanism or not, the thread initiation process serves as an entrée to locating cosmopolitanism within a digital tourism landscape. The poster may be interested at the onset as to how others feel about a particular topic or destination, and then defend or enter into robust debates online when the need arises. As Ong (2009) carefully denotes, cosmopolitanism is a highly iterative process with a long-term orientation. Therefore, through the thread initiation process, TT forums are a scaffold for digital tourism cosmopolitanism to be developed.

By reviewing all 216 trolling threads, 117 (54%) threads were initiated as trolling posts to attract participants, and 99 (46%) were hijacked by trollers. Emanating from this section is the potential that trolls are lurking throughout the TT forum waiting for an opportune moment to activate their deviant behaviour. Given the range of trolling incidents recorded, it is essential to further unpack circumstances related to troll encounters.

Troll encounters

This section scrutinises the troll encounters and its effect on digital tourism cosmopolitanism. Of interest to this research was the instance when the term 'troll' was initiated within the forum thread, and the circumstances that led to the poster or others being classified as a troll. The data revealed that trolling was identified across a range of discussions, and could appear anywhere at the onset of the thread, or at a later stage, as exemplified in the following exchanges of comments:

> Just wanted to ask how likely am I going to see lions\leopards\cheetahs if I visit a national park in Ethiopia. Like omo, mago, gambella or awash. Which one will i be likely to see one? (Keith)
>
> …
>
> PS I have a question. Do you troll other sites as well? (Bernard)
>
> Hi, I would like to go this summer to Ghana. Most of the time i would be in a volunteering camp, but i would also like to make alone some tourist visits throughout the country. Is it a socially safe country? The ISIS is there? Thanks! (Nancy)
>
> …
>
> The ISIS is there? Troll! (Harold)
>
> I know, I know, ancient cultures, sunny weather, great food, etc … but why should four young men (20 yrs) decide to go to Mexico? The main interests are drinking and women. Why Mexico and not Thailand for example? (Ben)
>
> …
>
> Here is the Original Poster's opening line from his Guatemalan post a few days ago:
>
> I'm working on a piece of FICTION. I think he and his friends should check into the TROLL MOTEL … … .. (Bryan)

These above-mentioned troll encounters occurred between exchanges of the thread initiator and others in the thread. Trolling was also reported among disputes arising between forum users that did not involve the original poster. In these incidents, there is a robust debate as to who is entitled to the 'right' perspective:

> Dear Bob, I really don't know how to help you for your message. It was so valuable and helpful that there are

only two words that come to my mind at the moment. The second is 'off.' (Amy)

The response I expected from someone not bright enough to plan ahead. (Bob)

Dear Bob, why do you respond if you have nothing (good) to say? (Amy)

To entertain the trolls, eh! (Bob)

Original Poster (OP), I think you are a troll. And if you are not a troll, I don't think you should be travelling. (Norleen)

Me (a troll)? Well thank you. I am just anxious for traveling about without family for the first time. Good for you that you know it all and are sure of yourself, I am not.

If you can't give any advice then don't reply to be honest. (Melissa)

Name calling doesn't make your post right ... Mountains are everywhere not only Mindanao. Your postings about Mindanao are generalization and can be summarized as all bull and speculation. I'm not a foreigner. I was born and raised in Cagayan de Oro. You've been to Mindanao 3 times and you see all of Mindanao is flooding and mudslide? For goodness sake Mindanao is many times bigger than Cebu and Bohol combined. Can't really fathom some postings that are speculation and generalization. (Billy)

Well, since I'm not the only poster on this thread who has questioned your intelligence (see link about Hemispheres) I don't think I need to continue a conversation with a troll like you. Good luck in your life. Ciao. (Eric)

As indicated in the exchanges above, the stance taken to identify and deal with trolling behaviour on TT forum is highly contextual. There did not appear to be any pattern as to when trolls were identified, nor trends that initiate the labelling of a troll. However, the data revealed some insights that provided interesting perspectives as to where trolling was more evident, and the timeframe in which they occurred.

As shown in Figure 2, most of the incidents of trolling are reported from TT forum threads relating to Central America. In taking a continental perspective, it is interesting to note that trolling occurred in the Americas on 82

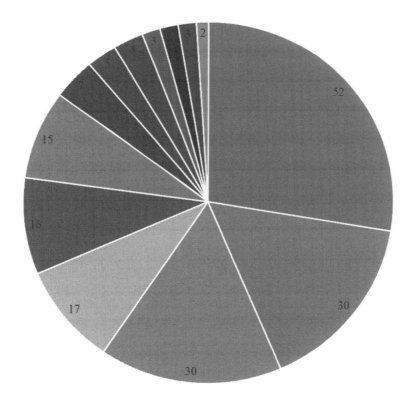

Figure 2. Geographical dispersion of trolling encounters by global regions.

out of the 188 TT threads (43.6%). The prominence of troll encounters in the Americas was hypothesised to be attributed to the larger volume of contents within this continent as compared to the rest of the world. However, as at the middle of October 2016, TT forums indicate that the proportion of threads across the continents in Figure 3 do not correspond to the patterns of trolling behaviour shown in Figure 2.

Despite having approximately just one-fifth of the threads globally, the prevalence of trolling behaviour in the North and Central America continent TT forums made it even more a surprising finding in light of tourism cosmopolitanism. However, one may be able to glean some insights from extant literature outside of tourism.

For instance, some scholars challenge cosmopolitan thinking and behaviours to be dominated by Western, postcolonial theories (see Kurasawa, 2004; Robbins, 1992), and thereby framing cosmopolitanism using a highly elitist and patriarchal perspective (Clarke, 2012). Such paradigms may essentially underpin many aspects of contemporary society, which according to McEwan and Sobre-Denton (2011), find its way into digital cosmopolitanism. Applied to the research, it was also evident in some of the TT threads that suggest that some of the trolling encounters were stereotyped to be about American identity and beliefs. Among the 164 threads mentioning the term America or the US, about a tenth of these contained stereotypical views that are best encapsulated in the following quotes:

> Americans do tend to be reluctant to pick up hitchhikers, due to too many stories of bad encounters. Even a girl with a backpack alone in the middle of the National Park. (Xavier)

> Probably you're also mixing up measles with malaria. Some people suspected of Ebola appeared to be malaria infected. Ok, facts and perception (indeed, the Fox-news-disease, a much more dangerous virus)... I can't help saying my first thought was: American. (Adam)

> I had always gotten top grades in my English classes but it wasn't until later reading well-written articles and books that I saw how we were not well taught... I shudder to think what today's students aren't learning ... and we do see examples here on TT often, and yes, from posters from the US. (Suzie)

> To be blunt, and assuming the original poster (OP) isn't a troll, this is the sort of rubbish Americans speak about everywhere that isn't America - and then you'll wonder why Americans are held in contempt by residents of other countries. (Heidi)

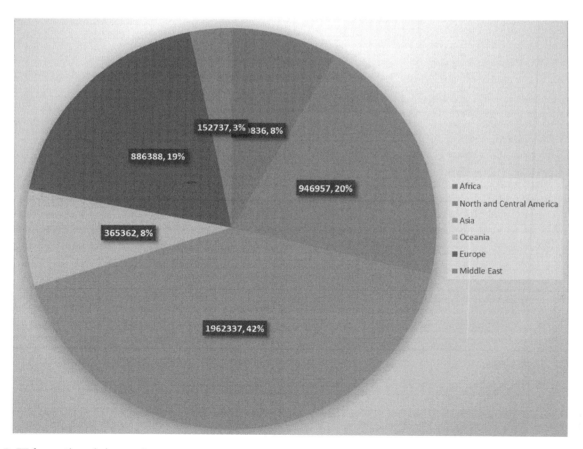

Figure 3. TT forum threads by continent.

These above-mentioned quotes insinuate the unsubstantiated perspective that cosmopolitanism should be delineated from a pro-American perspective, even in the context of digital tourism spaces. Such misconceptions should be challenged on the basis that trolling remains primarily fixated as attacks on persons (as indicated in Table 3), who may be physically located anywhere in the world. The research will subsequently turn to the effects of trolling encounters on re-orienting TT forum users and the effect on tourism cosmopolitanism.

Effect on tourism cosmopolitanism

Encounters with trolls on TT have had different effects related to the conceptualisation of digital tourism cosmopolitanism. The data revealed three main responses to trolling encounters. These are ignorance, rebuttal and a re-orientation to the proposed tourism experience. Table 4 denotes the frequency of these outcomes of trolling encounters by the original poster.

As seen in Table 4, the responses largely conform to the ethos of ignoring trolls on the TT forum, as one moderator carefully explains:

> I would like to remind people that part of the rules involve not being a troll, as well as not feeding the trolls. If you suspect someone is trolling, simply ignore them and report their post to the moderators for further action. (Ryan)

This stance on ignorance was evident in several troll encounters, with several TT forum users emphasising the need to avoid further engagements with identified trolls:

> Ignore poster #1. At the moment, he's the TT troll not to be indulged. (Wendy)

> Poster XYZ is an anti-Semitic troll and you'd better ignore him. Anyway, don't worry. Like everywhere in the world, there are some people with anti-Semitic world views but the majority of people are just fine and this should not stop you from visiting. (Heidi)

Most responses that have employed ignorance as a strategy lends a fresh perspective of the concept of digital cosmopolitanism. As discussed earlier in the paper, the digital environment facilitates opportunities to gain more knowledge through reading and engaging others in a space that can be more efficient and convenient. However, the affordances provided by such troll encounters casts a negative light to suggest that forum users are also likely to contend with others who may hold opposing views and thereby engage in some form of 'psychological warfare' (Workman, 2010). Such encounters go beyond what has been espoused as the principles of cosmopolitan studies as prescribed by Morais and Ogden (2011). So instead of becoming more globally aware and engaging with others to build up one's learning and knowledge, trolling fosters a hostile atmosphere that potentially leads to de-sensitising one's engagement on TT forums.

Such an outcome is also evident in responses that featured rebuttal within the forums. There were several exchanges on TT forum where several posters justified their perspective as the right and legitimate version:

> I am not a troll since I am human not a robot and the information I posted is true and useful. (Max)

> Me a troll? Well thank you. I am just anxious for traveling about without family for the first time. Good for you that you know it all and are sure of yourself, I am not. If you can't give any advice then don't reply to be honest. (Michelle)

> Don't get emotional by calling me a troll … The purpose of this thread is for any foreigners to give their opinion on this subject. It's my thread and therefore my rules. (Dylan)

The rebuttals epitomise the view that forums are a platform for individuals to seek homophilic outcomes. The desire appears to be for participants to get others to validate their choices and address some form of post-decision dissonance. Such exchanges raise further questions as to what digital tourism cosmopolitanism entails, and to what extent are forums enabling desired outcomes to be realised.

In dominant discourses on cosmopolitanism, rebuttals and debates are central to intellectual curiosity and have the potential to shift towards a more nuanced conceptualisation of cosmopolitan thinking and behaving (Beck, 2002; Heater, 2000; Osler & Starkey, 2003). However, in this instance, the debates appear to be raged around who is entitled to an opinion, and that those disagreements that have spiralled out of control are conveniently labelled 'troll'. A useful perspective of this unusual phenomenon may be discussed considering the nature of disputes in a face-to-face, versus an online context. In face-to-face conversations, disputes are centred around a small group of individuals (such as a meal at a table) where the protagonist or antagonist can easily change topic or focus attention elsewhere, especially if there are strong ties between them. It is highly unlikely that someone will be labelled a troll under such circumstances. For instance, Lynn commented that:

Table 4. Responses to trolling encounters and corresponding frequencies.

Response	Frequency
Ignorance	107
Rebuttal	98
Re-orientation	11

I don't think 'I've never been around the world before. What countries aren't safe?' is an ignorant or unreasonable thing to ask. Although to some degree I feel reassured that my question has been deemed ridiculous to some … I actually went to work today and asked why my question on this forum was ridiculed. 'Was it something I said?' In reply to this I got a whole bunch of wonderful stories that freaked the crap out of me.

In contrast, the online domain such as forums features an exchange of views among users who are likely to be strangers. The way the contents are drafted are therefore posted with greater anonymity, and far less accountability (Crawford, 2009; Scott & Orlikowski, 2012). Therefore, trolling encounters can be kept at arm's length without much repercussions to both the troll and the victim. Often these threads are shut down by administrators or moderators, though such deviant behaviour may resurface in another thread, as this paper has demonstrated.

While the findings may suggest that digital cosmopolitanism is of little consequence in altering one's global perspective, some TT threads reveal that, to a small extent, users may re-orientate themselves considering the trolling encounters:

> Alright people, original poster (OP) here. I have toned down my trip a bit. I guess it was too ambitious. It's now just 3 countries in 30 days. I realized that I would want to spend a lot of time in the Galapagos and wouldn't get any time in Panama … (Chris)

> Thanks XYZ and ABC for correcting me and sorry for my mistake. I wrongly assumed that a 'sight' is a place to be seen like you can see e.g. the Eiffel Tower or the Trevi Fountain just by looking at them. I now understand that a sight also means places that need to be properly visited because they are a complex and you can't just watch them while standing in front of it, like an archaeological site or even a museum. (Alvin)

> Thank you so much everyone!! I decided not to drive, and get a scooter or a bike while there :) The drive is too long, I don't want to get lost and my Spanish is at 3rd grade level … To everyone who responded, I am truly grateful :) (Sylvia)

> PS I'm not comparing ABC to another person in a third world country. You completely missed my point. And no thanks for dismissing my feelings. You shouldn't do that to any other person on this planet. My god. I'm done with this forum! (Alex)

The above-mentioned posts show that about 5% of the threads have stated that their encounters with trolls on TT forums have resulted in re-orienteering their position related to perceptions of the destination, their travel itineraries or even attitudes to the forum. Incidentally, the data revealed no apparent trends to suggest that trolling was limited to individual characteristics. Rather, trolls attacked random individuals for possessing specific attitudes or views about destinations and itineraries, such as in the case of James:

> Thanks for those who provided some useful information, I wasn't expect quite such a hostile response from a seemingly friendly message board … As I mentioned it is an idea that is being floated at the moment and therefore no, not much research has been done into the topic. I was hoping that this would be a forum where advice/guidance could be gathered.

These insights reveal that while users consider the use of forums such as TT convenient to solicit others' views about tourism, the digital medium opens a raft of other controversial issues that can gradually erode the inherent benefits of initiating cosmopolitan thinking and behaving. Collectively, the findings espouse some fundamental differences in terms of conceptualising cosmopolitanism within a tourism, pre-visit context. Condensing the outcomes from the research has led to the identification of some of these differences which support the beliefs underpinning digital tourism cosmopolitanism as presented in Table 1.

Conclusion, limitations and future studies

In conclusion, this paper had set out to explore how trolling encounters impacted digital tourism cosmopolitanism. The need for such a study to be conducted stemmed from a fast-evolving space where social media, and other digital spaces, are the initial lens in which tourists consume, interact and make decisions on their future tourism experiences (Hudson & Thal, 2013). Despite these developments, the nature of trolling as a deviant behaviour in tourism remains under-studied (Mkono, 2015). This research addresses these gaps in knowledge by extracting 216 trolling examples on *Lonely Planet Thorn Tree* forum to show how such incidents have impacted the nature of digital tourism cosmopolitanism.

The findings suggest that trolling is dismissed as an act of mischief and is consequently shut down by the site administrators or moderators once these have been reported. However, the insights reveal the tensions within virtual interactions that have provided a highly superficial, and myopic view of cosmopolitanism. The paper provides a nuanced understanding as to how trolling encounters are a (mis)alignment of tourism cosmopolitanism to date, and offers a fresh conceptualisation of the power that trolling can wield on the (un)intended outcomes to become better global citizens. While different scholars (e.g. Johnson, 2014; Salazar, 2010) envision tourism cosmopolitanism to be an elusive, utopian state that is always work in progress, trolling as a subset of digital cosmopolitan spaces may be considered a road block or a speed bump in this journey.

As with any study, this paper is not without its limitations. It was also noted that those engaging in the trolling discussions were tourists, with other stakeholders such as residents or destination management organisations (DMOs) almost non-existent. This is somewhat surprising, as other studies have alluded to the presence of other stakeholders in engaging tourists (current and potential) on social media platforms (Lange-Faria & Elliot, 2012; Milwood, Marchiori, & Zach, 2013; Munar, 2012). As such, the voices or insights from these stakeholders were absent in this study. However, it may be beyond the reach of some DMOs to interact with all social media contents. Nonetheless, the use of Artificial Intelligence or Chatbots could help to mitigate some of the problems posed by trolling (de-la-Pena-Sordo, Pastor-Lopez, Ugarte-Pedrero, Santos, & Bringas, 2016). Another limitation of the research was to utilise data from a solitary source of *Lonely Planet Thorn Tree* forums. Other forums and social media sites may offer a different perspective of trolling and their impacts on tourism cosmopolitanism.

Future studies could document the insights of other stakeholders in addressing trolling behaviour, as called for by Hays, Page, and Buhalis (2013). Additionally, further studies could feature the use of qualitative interviews conducted with both trolls and victims of trolls to shed light on more insights on this highly neglected area (Bishop, 2013). Another avenue for investigation is to compare how disputes are resolved between those that are flagged as trolling encounters online and others that involve face-to-face communication. Collectively, these paths present useful bases to advance the understanding of such deviant behaviour and their impacts on tourism cosmopolitanism.

Overall, this research has contributed to the need to understand trolling and how it interacts with tourism. This body of knowledge can serve to inform the extant literature surrounding digital spaces and tourism cosmopolitanism, and reiterates the need to unpack assumptions about trolling as a moderator of cosmopolitan thinking and behaving.

Disclosure statement

No potential conflict of interest was reported by the authors.

ORCID

Aaron Tham http://orcid.org/0000-0003-1408-392X
Mingzhong Wang http://orcid.org/0000-0002-6533-8104

References

de Albuquerque, K., & McElroy, J. (1999). Tourism and crime in the Caribbean. *Annals of Tourism Research*, *26*(4), 968–984.

Alderman, D. H., Benjamin, S. K., & Schneider, P. P. (2012). Transforming mount airy into Mayberry: Film-induced tourism as place-making. *Southeastern Geographer*, *52*(2), 212–239.

Aroldi, P., & Colombo, F. (2013). Questioning 'digital global generations'. A critical approach. *Northern Lights*, *11*(1), 175–190.

Arsal, I., Woosnam, K. M., Baldwin, E. D., & Backman, S. J. (2010). Residents as travel destination information providers: An online community perspective. *Journal of Travel Research*, *49*(4), 400–413.

Beck, U. (2002). The cosmopolitan society and its enemies. *Theory, Culture & Society*, *19*(1–2), 17–44.

Beck, U., & Sznaider, N. (2006). Unpacking cosmopolitanism for the social sciences: A research agenda. *The British Journal of Sociology*, *57*(1), 1–23.

Bianchi, R. V., & Stephenson, M. L. (2013). Deciphering tourism and citizenship in a globalized world. *Tourism Management*, *39*(December), 10–20.

Bishop, J. (2013). The effect of de-individuation of the internet troller on criminal procedure implementation: An interview with a hater. *International Journal of Cyber Criminology*, *7*(1), 28–48.

Bishop, J. (2014). Representations of 'trolls' in mass media communication: A review of media-texts and moral panics relating to 'internet trolling'. *International Journal of Web Based Communities*, *10*(1), 7–24.

Bjork, P., & Kauppinen-Raisanen, H. (2012). A netnographic examination of travelers' online discussions of risks. *Tourism Management Perspectives*, *2–3*(April–July), 65–71.

Boczkowski, P. J., & Siles, I. (2014). Steps toward cosmopolitanism in the study of media technologies. *Information, Communication & Society*, *17*(5), 560–571.

Brown, M. W. (1957). *The three billy goats gruff*. New York: Harcourt.

Buckels, E. E., Trapnell, P. D., & Paulhus, D. L. (2014). Trolls just want to have fun. *Personality and Individual Differences*, *67* (September), 97–102.

Bui, H. T., Wilkins, H. C., & Lee, Y. (2013). The social identities of Japanese backpackers. *Tourism Culture & Communication*, *13* (3), 147–159.

Calcutt, L., Woodward, I., & Skrbis, Z. (2009). Conceptualizing otherness: An exploration of the cosmopolitan schema. *Journal of Sociology*, *45*(2), 169–186.

Clarke, N. (2012). Actually existing comparative urbanism: Imitation and cosmopolitanism in north-south interurban partnerships. *Urban Geography*, *33*(6), 796–815.

Coghlan, A., & Gooch, M. (2011). Applying a transformative learning framework to volunteer tourism. *Journal of Sustainable Tourism, 19*(6), 713–728.

Cohen, S. A. (2011). Lifestyle backpackers: Backpacking as a way of life. *Annals of Tourism Research, 38*(4), 1535–1555.

Crawford, K. (2009). Following you: Disciplines of listening in social media. *Continuum, 23*(4), 525–535.

Cuffy, V., Tribe, J., & Airey, D. (2012). Lifelong learning for tourism. *Annals of Tourism Research, 39*(3), 1402–1424.

Delanty, G. (2006). The cosmopolitan imagination: Critical cosmopolitanism and social theory. *The British Journal of Sociology, 57*(1), 25–47.

Dippelreiter, B., Birgin, G. C., Pottler, M., Seidel, I., Berger, H., Dittenbach, M., & Pesenhofer, A. (2008). Online tourism communities on the path to web 2.0: An evaluation. *Information Technology & Tourism, 10*(4), 329–353.

Dubin, R. (1959). Deviant behavior and social structure: Continuities in social theory. *American Sociological Review, 24*(2), 147–164.

Dwivedi, M. (2009). Online destination image of India: A consumer based perspective. *International Journal of Contemporary Hospitality Management, 21*(2), 226–232.

Easton, S., & Wise, N. (2015). Online portrayals of volunteer tourism in Nepal: Exploring the communicated disparities between promotional and user-generated content. *Worldwide Hospitality and Tourism Themes, 7*(2), 141–158.

Falk, J. H., Ballantyne, R., Packer, J., & Benckendorff, P. (2012). Travel and learning: A neglected tourism research area. *Annals of Tourism Research, 39*(2), 908–927.

Filieri, R., & McLeay, F. (2014). e-WOM and accommodation: An analysis of the factors that influence travelers' adoption of information from online reviews. *Journal of Travel Research, 53*(1), 44–57.

Gibbs, J. P. (1966). Conceptions of deviant behavior: The old and the new. *The Pacific Sociological Review, 9*(1), 9–14.

Goode, L. (2010). Cultural citizenship online: The internet and digital culture. *Citizenship Studies, 14*(5), 527–542.

Govers, R., Go, F. M., & Kumar, K. (2007). Promoting tourism destination image. *Journal of Travel Research, 46*(1), 15–23.

Hall, C. M. (2011). On the mobility of tourism mobilities. *Current Issues in Tourism, 18*(1), 7–10.

Hannam, K., Butler, G., & Parris, C. M. (2014). Developments and key issues in tourism mobilities. *Annals of Tourism Research, 44*(January), 171–185.

Haralambopoulos, N., & Pizam, A. (1996). Perceived impacts of tourism: The case of Samos. *Annals of Tourism Research, 23*(3), 503–526.

Hardaker, C. (2010). Trolling in asynchronous computer-mediated communication: From user discussions to academic definitions. *Journal of Politeness Research, 6*(2), 215–242.

Hays, S., Page, S. J., & Buhalis, D. (2013). Social media as a destination marketing tool: Its use by national tourism organisations. *Current Issues in Tourism, 16*(3), 211–239.

Heater, D. (2000). Does cosmopolitan thinking have a future? *Review of International Studies, 26*(5), 179–197.

Hernandez-Mendez, J., Munoz-Leiva, F., & Sanchez-Fernandez, J. (2015). The influence of e-word-of-mouth on travel decision-making: Consumer profiles. *Current Issues in Tourism, 18*(11), 1001–1021.

Herring, S., Job-Sluder, K., Scheckler, R., & Barab, S. (2002). Searching for safety online: Managing 'trolling' in a feminist forum. *The Information Society, 18*(5), 371–384.

Hollinshead, K., Ateljevic, I., & Ali, N. (2009). Worldmaking agency – worldmaking authority: The sovereign constitutive role of tourism. *Tourism Geographies, 11*(4), 427–443.

Hopkinson, C. (2013). Trolling in online discussions: From provocation to community-building. *Brno Studies in English, 39*(1), 5–25.

Hudson, S., & Thal, K. (2013). The impact of social media on the consumer decision process: Implications for tourism marketing. *Journal of Travel & Tourism Marketing, 30*(1–2), 156–160.

Hwang, Y., Jani, D., & Jeong, H. K. (2013). Analyzing international tourists' functional information needs: A comparative analysis of inquiries in an on-line travel forum. *Journal of Business Research, 66*(6), 700–705.

Iaquinto, B. L. (2012). Backpacking in the internet age: Contextualizing the use of Lonely Planet guidebooks. *Tourism Recreation Research, 37*(2), 145–155.

Jacobsen, J. K. S., & Munar, A. M. (2012). Tourist information search and destination choice in a digital age. *Tourism Management Perspectives, 1*(January), 39–47.

Jane, E. A. (2014). You're a ugly, whorish, slut. *Feminist Media Studies, 14*(4), 531–546.

Johnson, P. (2010). De-constructing the cosmopolitan gaze. *Tourism and Hospitality Research, 10*(2), 79–92.

Johnson, P. C. (2014). Cultural literacy, cosmopolitanism and tourism research. *Annals of Tourism Research, 44*(January), 255–269.

Kozinets, R. V. (2002). The field behind the screen: Using netnography for marketing research in online communities. *Journal of Marketing Research, 39*(1), 61–72.

Kurasawa, F. (2004). A cosmopolitanism from below: Globalization and the creation of a solidarity without bounds. *European Journal of Sociology, 45*(2), 233–255.

Kusumasondjaja, S., Shanka, T., & Marchegiani, C. (2012). Credibility of online reviews and initial trust: The roles of reviewer's identity and review valence. *Journal of Vacation Marketing, 18*(3), 185–195.

Lange-Faria, W., & Elliot, S. (2012). Understanding the role of social media in destination marketing. *Tourismos, 7*(1), 193–211.

Lee, C. (2012). 'Have magic, will travel': Tourism and Harry Potter's United (Magical) Kingdom. *Tourist Studies, 12*(1), 52–69.

Leung, D., Law, R., van Hoof, H., & Buhalis, D. (2013). Social media in tourism and hospitality: A literature review. *Journal of Travel & Tourism Marketing, 30*(1–2), 3–22.

Lewis, S. C. (2015). Reciprocity as a key concept for social media and society. *Social Media + Society*, Apr–Jun, 1–2.

Lu, C. (2000). The one and many faces of cosmopolitanism. *Journal of Political Philosophy, 8*(2), 244–267.

Mack, R. W., Blose, J. E., & Pan, B. (2008). Believe it or not: Credibility of blogs in tourism. *Journal of Vacation Marketing, 14*(2), 133–144.

McCosker, A. (2014). Trolling as provocation: YouTube's agonistic publics. *Convergence, 20*(2), 201–217.

McEwan, B., & Sobre-Denton, M. (2011). Virtual cosmopolitanism: Constructing third cultures and transmitting social and cultural capital through social media. *Journal of International and Intercultural Communication, 4*(4), 252–258.

Mehari, K. R., Farrell, A. D., & Le, A. H. (2014). Cyberbullying among adolescents: Measures in search of a construct. *Psychology of Violence, 4*(4), 399–415.

Mills, J., & Green, B. (2013). Popular screen culture and digital communication technology in literacy learning: Toward a new pedagogy of cosmopolitanism. *Journal of Popular Film and Television, 41*(2), 109–116.

Milwood, P., Marchiori, E., & Zach, F. (2013). A comparison of social media adoption and use in different countries: The case of the United States and Switzerland. *Journal of Travel & Tourism Marketing, 30*(1–2), 165–168.

Mitchell, R. D. (1998). Learning through play and pleasure travel. Using play literature to enhance research into touristic learning. *Current Issues in Tourism, 1*(2), 176–188.

Mkono, M. (2011). The othering of food in touristic eatertainment: A netnography. *Tourist Studies, 11*(3), 253–270.

Mkono, M. (2015). 'Troll alert!': Provocation and harassment in tourism and hospitality social media. *Current Issues in Tourism.* doi:10.1080/13683500.2015.1106447

Molz, J. G. (2008). Global abode – Home and mobility in narratives of round-the-world travel. *Space and Culture, 11*(4), 325–342.

Morais, D. B., & Ogden, A. C. (2011). Initial development and validation of the global citizenship scale. *Journal of Studies in International Education, 15*(5), 445–466.

Munar, A. M. (2012). Social media strategies and destination management. *Scandinavian Journal of Hospitality and Tourism, 12*(2), 101–120.

Narayan, B. (2013). From everyday information behaviours to clickable solidarity in a place called social media. *Cosmopolitan Civil Societies Journal, 5*(3), 32–53.

Ong, B. S. (2012). The perceived influence of user reviews in the hospitality industry. *Journal of Hospitality Marketing & Management, 21*(5), 463–485.

Ong, J. C. (2009). The cosmopolitan continuum: Locating cosmopolitanism in media and cultural studies. *Media, Culture & Society, 31*(3), 449–466.

Ooi, N., & Laing, J. H. (2010). Backpacker tourism: Sustainable and purposeful? Investigating the overlap between backpacker tourism and volunteer tourism motivations. *Journal of Sustainable Tourism, 18*(2), 191–206.

Osler, A., & Starkey, H. (2003). Learning for cosmopolitan citizenship: Theoretical debates and young people's experiences. *Educational Review, 55*(3), 243–254.

Ossewaarde, M. (2007). Cosmopolitanism and the society of strangers. *Current Sociology, 55*(3), 367–388.

Papathanassis, A., & Knolle, F. (2011). Exploring the adoption and processing of online holiday reviews: A grounded theory approach. *Tourism Management, 32*(2), 215–224.

Paris, C. M., & Teye, V. (2010). Backpacker motivations: A travel career approach. *Journal of Hospitality Marketing & Management, 19*(3), 244–259.

Pearce, P. L., & Foster, F. (2007). A 'university of travel': Backpacker learning. *Tourism Management, 28*(5), 1285–1298.

de-la-Pena-Sordo, J., Pastor-Lopez, I., Ugarte-Pedrero, X., Santos, I., & Bringas, P. G. (2016). Anomaly-based user comments detection in social news websites using troll user comments as normality representation. *Logic Journal of the IPGL, 24*(6), 883–898.

Pritchard, A., Morgan, N., & Ateljevic, I. (2011). Hopeful tourism: A new transformative perspective. *Annals of Tourism Research, 38*(3), 941–963.

Robbins, B. (1992). Comparative cosmopolitanism. *Social Text, 31/32*, 169–186.

Rokka, J., & Moisander, J. (2009). Environmental dialogue in online communities: Negotiating ecological citizenship among global travellers. *International Journal of Consumer Studies, 33*(2), 199–205.

Roudometof, V. (2005). Transnationalism, cosmopolitanism and glocalization. *Current Sociology, 53*(1), 113–135.

Ryan, C., & Kinder, R. (1996). Sex, tourism and sex tourism: Fulfilling similar needs? *Tourism Management, 17*(7), 507–518.

Salazar, N. B. (2010). Tourism and cosmopolitanism: A view from below. *International Journal of Tourism Anthropology, 1*(1), 55–69.

Schmelzkopf, K. (2002). Interdisciplinary, participatory learning and the geography of tourism. *Journal of Geography in Higher Education, 26*(2), 181–195.

Schuckert, M., Liu, X., & Law, R. (2016). Insights into suspicious online ratings: Direct evidence from TripAdvisor. *Asia Pacific Journal of Tourism Research, 21*(3), 259–272.

Scott, S. V., & Orlikowski, W. J. (2012). Reconfiguring relations of accountability: Materialization of social media in the travel sector. *Accounting, Organizations and Society, 37*(1), 26–40.

Sin, H. L. (2009). Volunteer tourism – 'Involve me and I will learn'? *Annals of Tourism Research, 36*(3), 480–501.

Sluga, G., & Horne, J. (2010). Cosmopolitanism: Its pasts and practices. *Journal of World History, 21*(3), 369–374.

Smeaton, G. L., Josiam, B. M., & Dietrich, U. C. (1998). College students' binge drinking at a beach-front destination during spring break. *Journal of American College Health, 46*(6), 247–254.

Sobre-Denton, M. (2016). Virtual intercultural bridgework: Social media, virtual cosmopolitanism, and activist community-building. *New Media & Society, 18*(8), 1715–1731.

Sparks, B. A., Perkins, H. E., & Buckley, R. (2013). Online travel reviews as persuasive communication: The effects of content type, source, and certification logos on consumer behaviour. *Tourism Management, 39*(December), 1–9.

Stoner, K. R., Tarrant, M. A., Perry, L., Stoner, L., Wearing, S., & Lyons, K. (2014). Global citizenship as a learning outcome of educational travel. *Journal of Teaching in Travel & Tourism, 14*(2), 149–163.

Swain, M. B. (2009). The cosmopolitan hope of tourism: Critical action and worldmaking vistas. *Tourism Geographies, 11*(4), 505–525.

Tanford, S., & Montgomery, R. (2015). The effects of social influence and cognitive dissonance on travel purchase decisions. *Journal of Travel Research, 54*(5), 596–610.

Tavakoli, R., & Mura, P. (2015). 'Journeys in Second Life' – Iranian Muslim women's behaviour in virtual tourist destinations. *Tourism Management, 46*(February), 398–407.

Thompson, C. J., & Tambyah, S. K. (1999). Trying to be cosmopolitan. *Journal of Consumer Research, 26*(3), 214–241.

Uriely, N., & Belhassen, Y. (2005). Drugs and tourists' experiences. *Journal of Travel Research, 43*(3), 238–246.

Uriely, N., Ram, Y., & Malach-Pines, A. (2011). Psychoanalytic sociology of deviant tourist behaviour. *Annals of Tourism Research, 38*(3), 1051–1069.

Vasquez, C. (2011). Complaints online: The case of TripAdvisor. *Journal of Pragmatics, 43*(6), 1707–1717.

Vasudevan, L. M. (2014). Multimodal cosmopolitanism: Cultivating belonging in everyday moments with youth. *Curriculum Inquiry, 44*(January), 45–67.

Walsh, N., & Tucker, H. (2009). Tourism 'things': The travelling performance of the backpack. *Tourist Studies*, *9*(3), 223–239.

Workman, M. (2010). A behaviorist perspective on corporate harassment online: Validation of a theoretical model of psychological motives. *Computers & Security*, *29*(8), 831–839.

Wu, M., & Pearce, P. L. (2013). Appraising netnography: Towards insights about new markets in the digital tourist era. *Current Issues in Tourism*, *17*(5), 463–474.

Xiang, Z., & Gretzel, U. (2010). Role of social media in online travel information search. *Tourism Management*, *31*(2), 179–188.

Xie, P. F. (2004). Tourism field trip: Students' view of experiential learning. *Tourism Review International*, *8*(2), 102–111.

Yar, M. (2012). E-Crime 2.0: The criminological landscape of new social media. *Information & Communications Technology Law*, *21*(3), 207–219.

Young, C., Diep, M., & Drabble, S. (2006). Living with difference? The 'cosmopolitan city' and urban reimaging in Manchester, UK. *Urban Studies*, *43*(10), 1687–1714.

Zeng, B., & Gerritsen, R. (2014). What do we know about social media in tourism? A review. *Tourism Management Perspectives*, *10*(April), 27–36.

Index

Notes: Page numbers in *italics* refer to figures
Page numbers in **bold** refer to tables
Page numbers with 'n' refer to notes

Adey, P. 63
Adler, J. 16, 18
adulthood, transition from adolescence to 17, 18, 19, 69
agonistic public sphere 3, 7
Altbach, P. G. 19
altruistic behaviour, and social identification 16–17
American University 65, 66
Andreotti, V. 74
Appiah, K. A. 123
Arendt, Hannah 3, 4, 7–8, 9
Aristotle 4
Arnett, J. J. 18, 19
Aroldi, P. 136
Australia 14; education travel in 99–106; Endeavour Scholarships and Fellowships 99; National Strategy for International Education 15, 99; New Colombo Plan 15, 99, 102; outbound mobility programmes 15, 16

Bacalar Chico Marine Reserve and National Park (BCMRNP) 77, 79, 80
backpacking 136
Bagnoli, A. 17–18
Baillie Smith, M. 41
Ballantyne, R. 116
Baptista, João Afonso 50
Barcelona 40
Barnett, C. 25
Baumeister, R. F. 17
Beck, U. 17, 27, 28, 32
Belize *see* conservation volunteer tourism
beneficence, *vs.* benevolence 28
Benson, Angela 1
Benson, M. 63
Berlo, J. C. 129
'Big Society' 6
Bille, Mikkel 54
Binkley, Sam 90
Bishop, J. 134
Bissell, D. 63
Bjork, P. 135
Blue Ventures (BV) 76, 77, 79, 80–81
Brondo, K. V. 82
Brosnan, T. 73
Brown, Chris 10
Brown, G. 44
Brown, Lorraine 54
Buhalis, D. 143
Bush, George H. W. 87
Butcher, Jim 1, 2, 3, 25, 26, 87, 88, 110

Butler, Judith 27
Bynner, J. M. 17

Callon, M. 28
Cameron, J. D. 74, 113
Canhane (Mozambique), sensory experience in: community leader's household 53–54; developmentourism in 51–52; medicine man 54; responsibility/duty 55; school 55–56; shallow well 54–55; sight 54; stroll in 52–53, 57, 58; touch 53, 56; water supply system 56–57
care, and cosmopolitanism 26, 27–29
Casmir, F. L. 16
Cassidy, Jonathan 10
'celebrity-corporate-charity complex' 45
Chandler, D. C. 7
'choice biographies' 17
Chouliaraki, L. 28
citizen science 73–74, 75–76, 80, 81, 82; *see also* conservation volunteer tourism
citizenship 4, 14; outsourcing of 9; republican 3, 5, 7, 8, 11, 113; *see also* global citizenship
Clarke, N. 25
class, and global citizenship 38–39; disciplining of national citizen 43–45; global action 42–43; non-elite cosmopolitanisms and ongoing potentials of global citizenship 45–47; particularity and multiplicity of global citizenships 39–42
Classen, Constance 53, 54
Clemmons, David 1
Cloke, P. 25
Coghlan, A. 10, 25
Cohen, E. 99
Colombo, F. 136
colonialism, and volunteer tourism 5, 8, 45
consciousness-raising, and volunteer tourism 26
conservation volunteer tourism 73–74, 75, 81–83; collaborative programme design 82; knowledge divide 81–82; knowledge production 79–81; local residents 81, 82–83; research methods and study site 76–77, *78*; resources divide 82; transformative learning experience 82; volunteer skill-building and giving back through science 77–79
consumerism 25, 124
cosmopolitan empathy 24–25, 38, 46; cosmopolitanism, empathy and care 27–29; discourses of global citizenship 25–27; emotions and 27–29; humanitarian disasters and cosmopolitan empathy 28; Lacanian psychosocial theory 29–32; Lacanian tourism studies 31–32; natural disasters and 28; psychosocial reading of 32–33; space and 28
cosmopolitanism 1, 8–9, 40, 43, 47–48, 74, 113, 122–123; care and 26, 27–29; classed division in 41; and digital landscape of

tourism 133, **133**; and educational mobilities 63–64; impact of deviant online behaviour on 132–143; limits of 122–130; moral 50, 51, 52, 55, 57–58; non-elite 45–47; promotion, in Washington, DC 65–66; see also moral cosmopolitans; specific types
Côté, J. 17
Crabtree, R. D. 26
critical cosmopolitanism 27
critical global citizenship 74, 75, 81
critical reflection 3, 113, 115, 116
Crossley, Émilie 24
cultural capital, and internships 67–68, 70
cultural competency, and education travel 98–99, 102, 103–104, 105
cultural differences, and global citizenship 14
cultural identity 16
cultural knowledge 103, 105, 122, 123
cultural learning 105
cultural production 124
Curtin, S. 16

Datta, A. 41
Davey, Peter 54
Davidson, Joanna 124
de Albuquerque, K. 134
democracy, and global citizenship 6–8
democratic citizenship education 87, 91, 94
Desforges, L. 17
developmentourism 51–52, 57–58
development studies 15
deviant behaviour 133–134; see also online behaviour, deviant
Dickinson, J. E. 16
digital cosmopolitanism 132, 133, **133**, 138, 140
Diprose, K. 10, 15, 43
discursive psychology 29, 30
Dobson, A. 113
domestic volunteering 91, 93, 94
donations, microfinance tourism 117
Dredge, Dianne 109

education: global citizenship 4, 6, 9, 63, 86–94, 109–118; mobilities 63–64, 67–70; non-elite young people see class, and global citizenship; study abroad experience 13–19; travel 97–100; see also mediation, in education travel
elite cosmopolitanism 38, 41, 43
Elsrud, T. 17
embodiment 52, 53
Emerson, P. 30
emotions, and cosmopolitan empathy 27–29
empathy see cosmopolitan empathy
Enlightenment 54
environmental monitoring, knowledge production through 80
Erikson, E. 17
ethical consumerism 25, 124
ethical tourism 1, 4, 10, 27, 109, 110, 111; see also microfinance tourism (MFT)
ethics, in tourism 57
Everingham, P. 40
experience economy 14, 57
experiential learning 14–15, 74, 97, 98, 99, 114, 118; actions 117–118; concrete experience 114–115; cycle 114; dialogues 115–118; reflections 106, 115–118
extimacy 31, 34n2

facilitators-educators see mediation, in education travel
faculty see mediation, in education travel
Fahim, U. 16
Fairtrade 9
Faulkner, Jocelyn 72

Favero, Paolo 60n2
Filep, S. 73
Finnegan, Ruth 53
Foucault, Michel 87
Frändberg, L. 17–18, 67, 69
freedom 7–8
Frick, W. B. 18
'friendship politics' 46–47
Frosh, S. 29, 30
Fukuyama, F. 1
Fund for American Studies 65

Gagen, E. A. 45
gap year 4–5, 6, 10, 18, 26, 41
Geertz, C. 16
Geissler, P. 53
Generation NGO 5
'genuine cosmopolitanism' 52
Giddens, A. 1, 9, 17, 27
global charity 39, 43, 45
global citizens, nurturing of 100, 102–104
global citizenship 1, 3, 4, 9, 14, 73, 86, 87–88, 98, 133; and conservation volunteer tourism 79, 80, 81, 82–83; cultural differences and 14; democracy and 6–8; deprioritisation of nation state and 7; disciplining national citizens and 43–45; discourses 25–27; education and 4, 6, 9, 63, 86–94, 109–118; localism and 7; meanings of 101–102; mediation 104–106; see also mediation, in education travel; nurturing global citizens 102–104; particularity and multiplicity of 39–42; politics and 7–10; and power/democracy 6–8; promotion, in Washington, DC 65–67; science and 75; soft 74, 79, 80; social identity-based theory of 98; in study abroad experience 13–19; thick/thin 112–114, *114*, 116, 117; and volunteer tourism 4–8, 9–10, 74–75; see also class, and global citizenship
global citizenship, educating tourists about 109–110; creating personal encounters 114–115; critical 74, 75, 81; critical reflection 115, 116; data collection and analysis 112, **112**; gap in provider's perspective 110–111; goals of MFT providers 112–114; microfinance as new setting for tourism 115; microfinance tourism 111–112, **111**; perceptions of MFT providers 114–118; post-tour support 117–118; roles of tour guides 115–117; 'thin–thick' education continuum 112–114, *114*
global competency 98, 100
global educational travel (GET) see study abroad experience
global environmental citizenship 75
globalism 88
'globally oriented citizens' 7
Global South 5, 6, 8, 10, 25, 38
Gooch, M. 25
governmentality, volunteer tourism as 87
Grabowski, Simone 13
Grand Tour 13, 18, 19
Gray, Noella J. 72
Green, B. 132
Griffiths, M. 46
Gros, F. 57
Guatemala see Quetzaltenango (Guatemala) volunteer programme, knowledge exchange in

Hall, C. M. 26
Hall, E. T. 16
Han, J. H. J. 41
Hanley, Joanne 97
Hannam, Kevin 62, 63, 64
Hannerz, Ulf 52, 58
happiness 90–91
Harrison, L. 99
Hays, S. 143

Heath, S. 5
Heidegger, M. 53
Helvetas 51, 52
Hendrickson, C. 126
Heron, B. 116
Hibbert, J. F. 16
Höijer, B. 28
Hook, D. 30
human capital, and internships 68
humanitarian disasters, and cosmopolitan empathy 28

identity: and global educational travel 98–99, 102, 105; indigenous 124, 129; and internships 68; and knowledge production 82; and study abroad experience 16–17, 18, 19
indigenous knowledge see Quetzaltenango (Guatemala) volunteer programme, knowledge exchange in
'inside the beltway' 62, 63
intercultural competence, and study abroad experience 16
international service learning 26
Internet see online behaviour, deviant
internships 64; in Washington, DC 64–70
interpersonal relationships, in travel: effects on self 16
Iordache-Bryant, Iulia 86

Jacobsen, J. K. S. 137
Jakubiak, Cori 86
Jasanoff, S. 76, 82
Johnson, J. 65
Jones, A. 64
Judge, Ruth Cheung 38

Katzarska-Miller, I. 98
Kauppinen-Raisanen, H. 135
King, A. 18
Kingsbury, P. 31, 32
Kipp, Amy 72
knowledge exchange see Quetzaltenango (Guatemala) volunteer programme, knowledge exchange in
knowledge production 74, 75, 79–81
Knox, D. 64
Kolb, D. 114
Kozinets, R. V. 135
Kyriakidou, M. 28

labour market, and internships 67–68
Lacanian psychosocial theory 29–32; application to cosmopolitan empathy 32–33
Landon, Adam 13
Lattitude Global Volunteering 6
Lave, Rebecca 75
Law, J. 28
Leach, M. 76
learning-in-action 114
Lévinas, Emmanuel 53
life politics, volunteer tourism as 87
lifestyle mobility 63
lifestyle politics 9
Lisle, D. 27
localism, and global citizenship 7
Lonely Planet Thorn Tree (TT) 135–136
Lowenhaupt, Anna 88
Lyons, Kevin 13
Lyons, Kevin Daniel 97, 98, 100

MacCannell, D. 54
McCaslin, J. 62
McCosker, A. 133
McDonald, M. 5
McElroy, J. 134

McEwan, B. 140
Malach-Pines, A. 134
Malpass, A. 25
marginalised groups 91, 92
marine conservation see conservation volunteer tourism
Massey, D. 28
Mathers, K. 41
'mediated quasi-interactions' 28
mediation, in education travel 97–98, 106; Australian context 99–100; cultural competency and global citizenship 98–99; informal activities 106; meanings of global citizenship 101–102; mediating global citizenry 104–106; nurturing global citizens 102–104; pastoral care 106; research methodology 100–101; tour guide as mediator 99
Meeker, Alexandra 72
Merriman, P. 63
Mezirow, J. 114
microfinance tourism (MFT) 109, 111–112, 118; creating personal encounters 114–115; goals of providers for global citizenship education 112–114; as new setting 115; online global platform 117; perceptions of providers on global citizenship education 114–118; personal encounters in 114–115; post-tour support 117–118
Mills, J. 132
Minh-ha, T. T. 17
Mitchell, K. 28
Mkono, M. 134, 135, 136
mobilities, educational 63–64, 67–70
mobility burden 68
Moisander, J. 132
Montuori, A. 16
Morais, D. B. 141
moral cosmopolitans 50, 51, 52, 55, 57–58
moral economy, of volunteer tourism 27
moralisation, of tourism 110
morality 50–51
Mostafanezhad, Mary 1, 5, 8, 26, 28–29, 32
Mozambique see Canhane (Mozambique), sensory experience in
multiculturalism 14
Munar, A. M. 110, 137

Nance, M. 116, 117
Narayan, B. 132
national citizens, disciplining: and global citizenship 43–45
national sovereignty 8
natural disasters, and cosmopolitan empathy 28
Neil, J. 25
Nelson, Rebecca L. 122
neocolonialism 5, 8, 40, 41, 44
neoliberalism 5, 40, 41, 82, 86, 90, 117, 127
'neoliberal science regime' 75
netnography 135
networked cultures 132
non-elite young people see class, and global citizenship

Obama, Barack 87
Ogden, A. C. 141
Ong, J. C. 138
online behaviour, deviant 132–133; conceptual framework of research *135*; effect of trolling on tourism cosmopolitanism 141–142, **141**; limitations and future studies 142–143; research method 135–136; thread initiation 137–138; troll categorisation 137, **137**; troll encounters 138–141, *139*, *140*; see also trolling
Osgood Centre 65
Ossewaarde, M. 138
Other, the 53, 55; objectification of 5; Third World 26, 32–33
outbound mobility programmes 15–16, 97, 100

INDEX

Packer, J. 116
Page, S. J. 143
Paine, Thomas 8
Palacios, C. 5
Panagia, Davide 59
Parekh, Bikhu 6–7, 8
Parilla, J. 62
participant observation, data from 42, 77, 89
'partition of the sensible' 59
Pearce, P. L. 10, 135
Perlin, R. 67, 68
Phi, Giang Thi 109
Platform2 43
'pleasure periphery' 31
Ploner, J. 67
Pogge, T. 52
polis 4
politics: friendship 46–47; and global citizenship 7–10; life 87; political power of tourists 110, 112, 113; in sensory experience of tourists 58–59
Ponting, T. 5
popular humanitarianism 38, 40, 42
post-tour support, microfinance tourism 117–118
Potts, D. 99
power: and global citizenship 6–8; political power of tourists 110, 112, 113; soft 102
Prince, R. 53
'progressive' cosmopolitanism 27
'psychosocial studies' 29
psychosocial theory *see* Lacanian psychosocial theory
public realm 7
purposive sampling 64

Quetzaltenango (Guatemala) volunteer programme, knowledge exchange in 122–123; development work 127–128; entanglement with other world views 123; ethnic identity 129; inventory and sales 127; knowledge as resource 124–125; learning about Mayan culture 127; Mayan heritage 126–127; research methods 125; strategic planning and project management practices 127; threat of knowledge theft 128–129; tourists as sources of information 125–128; weavers as sources of information 128–130; weaving museum 126

Raleigh International 6
Ram, Y. 134
Rancière, Jacques 59
Ravensbergen, Sarah 72
Raymond, E. M. 26
Reau, B. 18
reflective journals 105
reflective learning 15
Reid, Sacha 109
Reisberg, L. 19
republican citizenship 3, 5, 7, 8, 11, 113
resources management 104
'responsibilisation of poverty' 44
Reysen, S. 98
rites of passage 13, 18–19
Rock, J. 73
Rokka, J. 132
Romania, volunteer tourism in 86–87; bringing smiles 90–91; interaction of volunteers with residents 91–92; language barriers 91–92; listening to others 91–93; research methods 88–90; and self 93–94; working with children 92
Rumbley, L. E. 19

São Paulo Turismo (SPTuris) 57
Schattle, H. 14
Schön, D. A. 106, 114

Schubert, Felix 62
Schweinsberg, S. 100
science: citizen 73–74, 75–76, 80, 81, 82; giving back through 77–79; and global citizenship 75; neoliberal regime 75
science and technology studies (STS) 76
Scoones, I. 76
Scouts (UK) 43
self: aspirational 44–45; and global educational travel 98; and study abroad experience 16–17; and volunteer tourism 25, 93–94
Senator Paul Simon Study Abroad Program Act (US) 15
sense of worth, from volunteer tourism 47
'sensescapes' 57
sensory experience, of tourists 50; politics 58–59; pursuit of moral cosmopolitanism 57–58; *see also* Canhane (Mozambique), sensory experience in
sensory marketing 57
Sheller, M. 63
short-term mobility programmes (STMPs) *see* mediation, in education travel
Silk, J. 28
Simpson, K. 5, 26
Sin, H. L. 26
Sinervo, A. 27
skill-building, volunteer 77–79
Skrbis, Z. 28
Smith, D. M. 28
Smith, Peter 1, 25, 26, 87, 88
Sobre-Denton, M. 140
social contract 4
social identity-based theory of global citizenship 98
social identity theory 16–17
social inclusion 91
social media *see* online behaviour, deviant
social networks, and internships 66, 69
Socrates 109
soft global citizenship 74, 79, 80
soft power 102
solidarity 8, 46
Sørensen, Tim 54
space: and cosmopolitan empathy 28; psychoanalytic 31
Standish, Alex 7, 9, 63
Starr-Glass, D. 19
Stoner, K. R. 13, 97, 100
Stranger, Mark 55
strategic cosmopolitanism 38, 41, 44
study abroad experience 13–14; global educational travel 14–16; self-exploration and identity 16–17; youth transformation and rites of passage 17–19
sub-Saharan Africa *see* class, and global citizenship
Sussman, N. M. 16
Sutherland, L. A. 116
Sweeney, Trip 111
'symbolic growth experience' 18
Symbolic Order 30

Tarrant, Michael 13
Tarrant, M. A. 14, 100
Taylor, R. 19
'technology of the self' 45
textual analysis 64
Tham, Aaron 132
thematic analysis 112
thick global citizenship 112–114, *114*, 116, 117
Thien, D. 28
thin global citizenship 112–114, *114*, 117
Thomson, R. 19
Tønder, L. 59
tour guides: education travel 98, 99; microfinance tourism 115–117

INDEX

tourism *see specific types*
tourism studies: critical turn 130; Lacanian 31–32
tourists: compassion of 112–113; desire of 31, 32, 33, 46; fantasy of 31, 32, 33; political power of 110, 112, 113; *see also* global citizenship, educating tourists about; sensory experience, of tourists
Towner, J. 18
transformative learning 25, 82, 114
transience, and internships 68–70
TripAdvisor 135, 136
trolling 133, 134–135; categories 137, **137**; debates 138–139, 141; effect on tourism cosmopolitanism 141–142, **141**; encounters 138–141, *139*, *140*; ignorance 141; rebuttal 141–142; re-orientation 142; thread initiation 137–138; typology **134**
Trujilo, J. R. 62
Trump, Donald 62
Tucker, H. 29

United Kingdom (UK) 6; Department for Business, Innovation and Skills 6; International Citizen Service 6; Platform2 43; *see also* class, and global citizenship
United States (US) 6, 45; Lincoln Commission 15, 99
United States Agency for International Development (USAID) 51
United States Peace Corps 6
University of Newcastle (UON) 100
Uriely, N. 134

van Gennep, A. 13, 18
virtue, and volunteer tourism 47
Visiting Friends and Relatives (VFR) 69
Voluntary Service Overseas (VSO) 6
volunteer tourism 1–2, 3, 73; and citizen science 75–76; classed experiences *see* class, and global citizenship; colonialism and 5, 8, 45; consciousness-raising and 26; and conservation *see* conservation volunteer tourism; and cosmopolitan empathy *see* cosmopolitan empathy; cosmopolitanism 8–9; criticisms 26, 40; and global citizenship 4–8, 74–75; as governmentality 87; as life politics 87; moral economy of 27; outsourcing of citizenship to the globe 9–10; and power/democracy 6–8; pragmatic politics 9; in Romania *see* Romania, volunteer tourism in; sense of worth from 47; trips as shock therapy 44; and virtue 47
voluntourism *see* volunteer tourism
Vrasti, Wanda 1, 5, 8

Wang, Mingzhong 132
Ward, C. 16
Washington, DC, study-internship programmes 62–63; building competitive cultural capital 67–68; internship industry 64–65; promoting cosmopolitanism and global citizenship in 65–67; student mobilities 67; transience 68–70
Washington Center 65, 66, 67
Washington Internship Institute 65
Washington Semester Programmes (WSPs) 63, 64–65, 66, 67–68, 69
Wearing, Stephen 1, 5, 13, 25, 100
Weiner, A. 124
Whitford, Michelle 109
world state 7
Wu, M. 135
Wynne, B. 76

Yanapuma Foundation 6
Young, Tamara 97
youth movements 43
youth transformation *see* study abroad experience

Žižek, S. 31
Zuckerman, Martin 58